T0013664

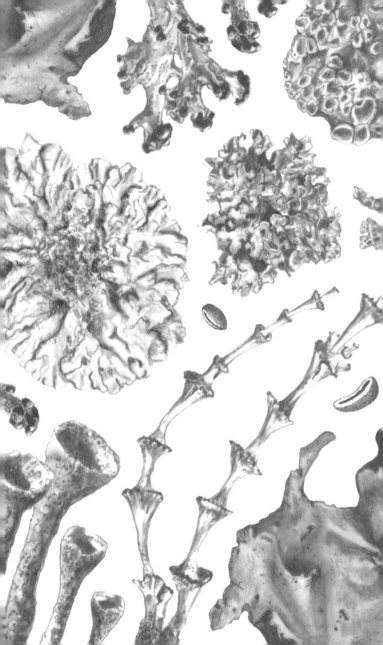

Lichenpedia

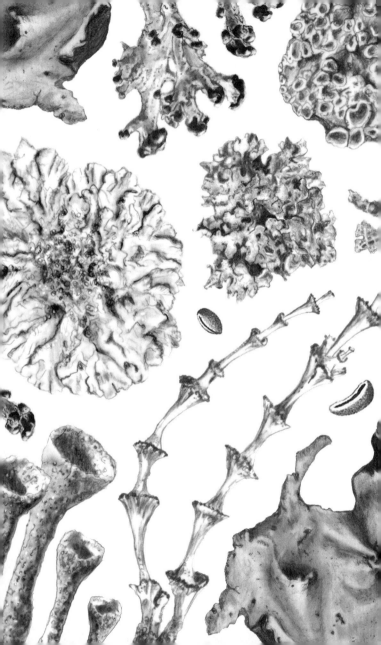

Lichenpedia

A Brief Compendium

Kay Hurley

Illustrations by Susan Adele Edwards

PRINCETON UNIVERSITY PRESS
Princeton & Oxford

Published by Princeton University Press
41 William Street, Princeton, New Jersey 08540
99 Banbury Road, Oxford OX2 6JX

press.princeton.edu

ISBN 9780691239903
ISBN (e-book) 9780691239897

British Library Cataloging-in-Publication Data is available

Editorial: Robert Kirk and Megan Mendonça
Production Editorial: Mark Bellis
Text and Cover Design: Chris Ferrante
Production: Steve Sears
Publicity: Matthew Taylor and Caitlyn Robson
Copyeditor: Cynthia Buck

Cover, endpaper, and text illustrations by Susan Adele Edwards

This book has been composed in Plantin, Futura, and Windsor

Printed on acid-free paper. ∞

Printed in China

10 9 8 7 6 5 4 3 2 1

*to Pliny the Elder,
compiler of the first encyclopedia,
which he wrote "for the masses,
for the farmers and workers, and to interest
people in their leisure time"*

Preamble

"*To see a World in a Grain of Sand . . .*"

—WILLIAM BLAKE, "AUGURIES OF INNOCENCE"

"*. . . is a puzzlement.*"

—OSCAR HAMMERSTEIN, *THE KING AND I*

Lichens are weird. Wonderfully weird.

The definition of a lichen as a dual organism made up of a fungus and an alga caused a huge furor in the nineteenth century. Many thought it was just plain wrong—how can one thing be two things? Some thought the fungus must be a parasite on the alga. Others thought they each derived benefit. These thoughts gave rise to a new word, "symbiosis," an umbrella term meaning "living together." There's lots of ways to live together. Parasitism is at one end of the spectrum, but if both parties benefit, as with lichens, the relationship is at the other end, the polar opposite of parasitism, and is called mutualism. A classic example of mutualism is that of bees with flowers. The bees get nectar and pollen to eat, and the plant gets the pollination services necessary for its reproduction. Other examples are coral with algae, and humans with their gut bacteria. The partners in these relationships definitely benefit one another, while retaining their own very distinctive identities. The huge difference with lichens is that the result of the partnership is a new entity not remotely like its component parts. The partners have gone beyond symbiosis.

We the lichenizing fungi of the world, in Order to form a more perfect Union, ensure domestic Tranquility, provide for the common Defence, promote the general Welfare, and secure the Blessings of Long Life to ourselves and our Photobionts, do ordain and establish the New State of Ultra Symbiosis.

The dual-organism definition was widely accepted by the start of the twentieth century and held until the twenty-first century. The duality was always of a fungus (called the mycobiont) and at least one photobiont (something capable of photosynthesis). The photobiont was usually an alga but sometimes a cyanobacterium and sometimes both. Schoolchildren, if they were taught at all about lichens, learned that Freddy Fungus and Alice Alga took a "lichen" to each other. This is the only knowledge of lichens that some adults retain today. We now know that there's more to the story, that our knowledge, as always, was incomplete. At least two different fungi are involved. In fact, as the University of Alberta lichenologist Toby Spribille, in a conversation with Merlin Sheldrake, author of *Entangled Life*, said, "We have yet to find any lichen that matches the traditional definition of one fungus and one alga."

Fungi cannot make their own food. Like us, they have found various ways to get it. The type of fungi that become lichens envelop photobionts (food makers) and give them protection and places to live where they would be unable to live on their own. The photobionts make enough food for both themselves and the fungi, but they are captives. At the risk of going down the rabbit hole of anthropomorphism, the photobionts seem to be happy prisoners with Stockholm syndrome. On the other hand, it could be that they "asked" the fungus to build them a home and happily pay rent in the form of food. Either way, both parties have made it work. The role of the other fungi present is still a puzzle—as are lichens themselves. They're not dual organisms, nor even triple, and they also have bacteria; they are actually tiny ecosystems—"a world in a grain of sand."

The seven wonders of the natural world include the Grand Canyon, Mount Everest, Victoria Falls, and the Aurora Borealis. What an impossible task to choose only seven! Better to ask, what in the natural world is not a wonder? Because, for example, lichens! Most people in the world will never get to see firsthand any of the seven grandiose wonders, but everyone can be awed by lichens. They're present year-round, and they're everywhere, hiding in plain view all around us—on tree trunks, rocks, the ground, leaves, and many artificial materials. On a lichen-scouting outing on Mount Washington in New Hampshire, Alan Fryday, master of alpine lichens, somewhat jokingly asked if anyone could find a rock with no lichens on it. But you don't have to go to the mountains, live in the country, or visit conservation land. People in cities can find lichen on trees in the supermarket parking lot, on steel fence railings, on monuments, on building foundations, and even on the pavement. Some are growing on the concrete on my front door step. I barely have to leave the house.

Lichens are well-behaved citizens of the environment with much to offer. They are nature's pioneers, colonizing bare rock and starting the process of soil formation. They grow on disturbed soil, hold it against erosion, and allow plants to get a foothold. They fix nitrogen. They provide shelter for lots of animals, from tiny tardigrades to birds and squirrels that use them for nesting material. They provide food for many animals, from invertebrates to large mammals like moose and reindeer. They live long lives and sequester carbon. They are innovative chemists, producing more than 1,000 chemical compounds, most of which are not found anywhere else in nature. Humans, less well-behaved citizens of the environment, use lichens for dyeing, brewing, and monitoring air pollution; we use lichens for medicine, either directly or as a starting point for synthesized pharmaceuticals, for measuring the age of rock surfaces, occasionally for food (especially in South Asian and East Asian cuisine), and for artistic inspiration.

Every branch of science has key people who advance the body of knowledge through their discoveries and by sharing knowledge in their writing and teaching. Some are ancient, like Theophrastus and Pliny the Elder; some are less ancient, like Schwendener and Tuckerman; and some are alive, productive, and largely obscure—like algae in a lichen. I have included in *Lichenpedia* a few of their stories and discoveries.

Every branch of science has its own specialized language, and lichenology is no exception. It's hard to talk about lichens unless you are familiar with at least a few of the hundreds of terms used by lichenologists. I explain the three main growth forms: foliose, crustose, and fruticose. The body of a lichen, the thing you see regardless of the growth form, is called a thallus. Fungi that don't form lichens are seldom visible. They live in the ground or inside a tree, and we discover their presence only when they produce fruiting bodies, such as a mushroom. Fungi that form lichens, on the other hand, are highly visible, whether fruiting or not. The various parts of a lichen responsible for reproduction, like apothecia, soredia, and isidia, are also useful terms, so I have included them too. There is also rhythm and music, if not romance, in the sound of the words themselves. Take them to heart. How could you not love a foliose thallus with lecanorine apothecia sprinkled with soredia looking like a pepperoni pizza dusted with Romano? "Processed meat on baked dough with cheese" doesn't do it. Good morning, Soredia. Rise and shine, Apothecia.

Much of the technical language and nomenclature is drawn from Greek and Latin. If reviving some of your high school Latin doesn't give you joy, try your hand at rap or parody and go to town with genus names. In the "Song of *Ramalina*—Ding Dong," I would leave my flattened branches by the shores of Gitche Gumee and search for *Rinodina*, hand in hand with Hiawatha hunting for some *Opegrapha*.

Lichens have so much to offer. Besides the many mysteries they hold for professional lichenologists, lichens are engaging and intriguing for amateur naturalists and could become a lifelong hobby. (The only downside is that you can't hike with your friends anymore—you go too slowly and stop too often.) Lichens take you outdoors and put you in touch with the natural world in a new way. Overlooked by so many, they offer kinship to those who love underdogs. They are extremophiles. They survive without water better than a camel, and in Earth orbit without a space suit; they live in both deserts and Antarctica. Lichens are carbon sinks and could transform a world challenged by deforestation and climate change. Their exquisite detail and variety in form and color are both a challenge and a temptation for artists, and poets, writers, and philosophers find in them a metaphor for everything.

My wish for all readers is that you find pleasure in lichens, in one way or another. Nobody captured it better than the writer and explorer Lawrence Millman did in his poem "Lichens" (published in his 2000 collection *Northern Latitudes*), "May the gods of the tundra grant me lichen until I become lichen myself."

Graphis scripta

Acharius, Erik (1757–1819)

Thank you, Erik Acharius, for paying attention to lichens. This man established the first taxonomic classification of lichens following the Linnaean binomial system and came to be known as the father of lichenology. Linnaeus had been happy to lump a handful of lichens into the single genus *Lichen*. Acharius created 40 genera to accommodate over 3,300 lichens.

Before Acharius got into that work, he studied natural history and medicine under Carl Linnaeus at Uppsala University, Sweden, and was the last student to defend a dissertation before him. Acharius was twenty-one years old when Linnaeus died. He practiced medicine for a short time and eventually became a provincial medical officer for the area around Vadstena, East Gotland. One wonders if the job was not demanding, leaving him a lot of free time to spend on lichens. In 1798, after he'd been at it for less than ten years, he published a *Lichenography of Sweden* (*Lichenographiae Svecicae Prodromus*), detailing all the lichen species known in Sweden—a tough assignment even without a day job. This publication was the first to use the Linnaean binomial system for lichens. Another Swedish lichenologist, Olof Swartz, who was also a believer in the Linnaean system, exchanged hundreds of letters with Acharius and is thought to have heavily influenced his work. Swartz could be the unacknowledged stepfather of lichenology.

The binomial system is easier to appreciate when compared to earlier attempts to name lichens. For example, Johann Dillenius contributed much to the body of knowledge with drawings and descriptions, but he was not so good with names. His common name for a certain lichen was Strange-character'd Lichen, and the Latin name was a string of descriptors: *Lichenoides crusta tenuissima peregrinis velut litteris inscripta*. Linnaeus placed the same lichen in his one and only genus for lichens and reduced the six-word mini-novel to one word, "scripta," giving us *Lichen scripta*. Acharius called it the Common Script lichen, classified it as graphoid, and gave it the binomial *Graphis scripta*—all of which stand today. He used the shape, color, and location of both vegetative and sexual reproduction structures to support his classification logic and decisions.

We can deduce that Acharius had the personality of an optimist. Why else would he publish a book with the title *A Method by Which Everyone Can Identify Lichens*

(1803)? Everyone! He published two more important works: *A Universal Lichenography* and *Taxonomical Arrangement of Lichens*.

He died at age sixty-one of an apoplectic stroke suffered at his microscope after spending the morning studying a collection of lichens from Spain. A coincidence? No one knows. Lichen identification can sometimes be challenging, even frustrating, but seldom to the point of apoplexy.

The highest honor awarded in lichenology is the Acharius Medal. Granted by the International Association for Lichenology, an organization that promotes the study and conservation of lichens, it recognizes outstanding lifetime contributions to the field.

See also Linnaeus.

Air Pollution

Lichens are known for their ability to endure some of the harshest conditions on Earth and even in space, but this tough group of organisms includes many species with an Achilles' heel. Air pollutants such as sulfur dioxide, nitrogen oxides, fluorides, heavy metals, and acid rain kill them.

Three main issues make lichens vulnerable. First, they get the minerals they need directly from the air and from rainwater on their surface. Because these minerals were once present in very low concentrations, lichens developed efficient methods to absorb them and historically had no need to discriminate between good and bad. Second, lichens live long enough that minerals can accumulate in them to a toxic level—and when the concentration in the air or water is high, toxicity is reached quickly.

Third, lichen health depends on the balance between the symbiotic partners. If something causes one to get too far ahead or to fall too far behind, the partnership breaks down. Irreconcilable differences.

The Industrial Revolution started in England around 1760, but it wasn't until 1807 that the link between air quality and lichen health was first noted. France began its Industrial Revolution in the mid-nineteenth century, and by the end of the century the lichen species previously documented in the Luxembourg Gardens in Paris were all gone. A similar situation in Stockholm in 1926 gave rise to the term "lichen desert"—an area with no lichens at all. Surveys in New York City showed that 191 species documented in 1824 had dropped to 51 by 1914 and were down to five by 1968. In the early 1900s, these declines were thought to be caused by soot. By 1970, sulfur dioxide was identified as the major culprit.

Lichen's sensitivity to air pollutants, especially sulfur dioxide, enables us to gauge air quality by monitoring lichen diversity. Diversity is the key factor, because not all lichens are equally sensitive. By and large, foliose and filamentous species are more vulnerable than crustose species. Poor air quality is indicated in areas with only pollutant-resistant lichens.

In the summer of 1971, 15,000 UK schoolchildren equipped with a basic kit from the Advisory Centre for Education documented lichen species across the entire country. The simple protocol they used resulted in a nationwide map of polluted and unpolluted areas to serve as a baseline for later surveys. Although more sophisticated protocols are now widely used by government agencies such as the US Forest Service, the US National

Park Service, and the Canadian Environmental Monitoring and Assessing Network (EMAN), a role for citizen science projects remains. Conservation commissions, park managers, and local land trusts have a way, courtesy of lichens, to understand more deeply the health of the habitats they administer.

The passage of clean air acts around the world has led to cleaner air and a corresponding recovery in pollution-sensitive lichens. By 2021, the New York City count was up to 106 species. Significantly, one of those species was a beard lichen (*Usnea*), notable because lichens of this genus are very vulnerable to pollution, and none had been recorded since 1824.

Alcohol

The photobionts in lichens contain starch. Where there's starch and humans, there's a strong possibility there will be alcohol production. Lichens are not the preferred source of starch for alcohol, especially where there are large annual crops of grain or potatoes, but they were tried in Sweden and Russia, where lichens are plentiful.

George Llano, the twentieth-century polar explorer and lichenologist, included a history of alcohol distillation from lichens in a 1948 paper written during his PhD studies at Harvard, "Economic Uses of Lichens." His paper was recognized as important and subsequently published in several journals. Llano reported that Swedes in the nineteenth century saw lichen alcohol as an alternative to grain alcohol. They proved the feasibility of the idea, claiming a yield of five liters (5.3 quarts) of 50 percent alcohol from twenty pounds of lichen. By 1893, lichen brandy had become a large industry in

Sweden—too large for its own good as the rate of lichen harvesting outpaced lichen regrowth. The industry declined. Meanwhile, Russians were taking advantage of the Swedes' proof of concept. At the Russian Industrial Exhibition of 1872, several distillers exhibited samples of lichen spirits, which were well received by visitors from France and England.

Lichens have also played a role in the brewing of beer. A Siberian monastery used lungwort lichen, *Lobaria pulmonaria*, as a substitute for hops. The beer was served to travelers, who detected a peculiar bitterness and found it to be highly intoxicating. Lichens are well known for their bitterness and were probably responsible for the strange taste but not for the intoxication level.

The Tarahumar people of Mexico, famous for their long-distance running, are less famous for their use of lichens (*Usnea* species) in making their traditional fermented corn beverage, *tesgüino*. This alcoholic drink has an important place in the culture of the Tarahumara. How long before craft beers, which are growing in significance in our own culture, include a lichen-flavored version?

See also glucose.

Algae

The first thing that comes to mind when people think of algae is probably not lichens. The usual suspects are pond scum or seaweed, both of which are in fact algae, a catchall term that covers a highly variable group of "things." Some of them are related to land plants and some of them are not. Algae, like mosses, liverworts, and hornworts, don't have stems, roots, leaves, flowers, seeds, or berries. The one thing they have in common with each other and with land plants is that they photosynthesize.

"Algae" as a catchall term no longer encompasses the three traditional groups: blue-green algae, macroalgae (or seaweeds), and single-celled microalgae. Blue-green algae are now recognized as cyanobacteria, which leaves just the macros and the micros. Species of macroalgae are a nutritious food source for humans. In Japan and Korea, they make up about 10 percent of the diet. Microalgae include diatoms, dinoflagellates, and green algae. Some of them form phytoplankton. Microalgae are massively significant for two reasons: they are the base of the aquatic food chain around the world, and as a by-product of photosynthesis, they produce about 50 percent of the oxygen we breathe. It's time to stop calling them scum.

Microalgae form important symbiotic relationships. For example, corals depend on dinoflagellates, and about 25 percent of the world's fish species depend on coral reefs. Spotted salamander embryos depend on algae for the oxygen their egg mass needs to survive, and 90 percent of all lichen species would not exist without green algae to make their food. (The remaining 10 percent depend on cyanobacteria.)

Not all green algae participate in lichen symbiosis. The most common species that do are in the genera *Trebouxia* and *Trentepohlia*. Of those, *Trebouxia* is mostly found in combination as a lichen. *Trentepohlia*, on the other hand, is frequently found free-living. Despite being a green alga, it has pigments that make it look orange or rust-colored. You may have seen large colonies of it in bark crevices on trees or on limestone, usually on the shady side. Of these two alga species, *Trebouxia* is the photobiont in roughly 70 percent of lichens, which is odd. You may wonder how a germinating fungal spore would ever encounter any *Trebouxia* to capture. One theory is that the germinating spore finds fragments of existing lichens with the alga in its combined form and is somehow able to acquire it.

Most of the algae in lichens have not been identified beyond the genus level. The problem is that they are changed a lot by their life inside the thallus. To be sure of their identity, they used to be isolated and grown in culture; nowadays DNA sequence data are becoming standard means of identification.

Antarctica

This dry, windy, icy continent was home to 350 species of lichens in 2001. A recent estimate has pushed that up to closer to 450. The additions are species not previously known in Antarctica and also species new to science.

The adjective "Antarctic" describes more than just the continent itself. The Antarctic *Circle* is approximately 66.5° south of the equator. Its position is based on astronomical calculations and excludes bits of the continent at its northern edges. The Antarctic *territory*, a term used in ecological studies, includes the whole continent and all the land south of latitude 60°. It has two biogeographical zones: the "continental zone" is the continent itself plus islands surrounded by the ice shelf; the west coast of the Antarctic peninsula is included in the "maritime zone," along with all the other islands below latitude 60° such as South Shetlands, South Sandwich, and South Orkneys.

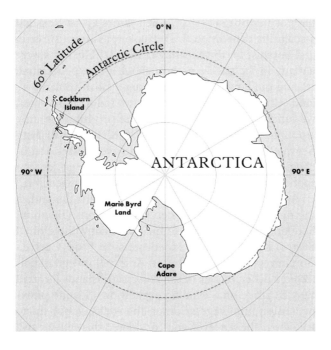

The maritime zone is the warmer and more humid of the two biogeographical zones, with obvious biological implications. The two Antarctic vascular plants are known only in the maritime zone. The first report of an Antarctic lichen was also from the maritime zone when Captain William Napier of the US Sealing Expedition of 1820–1821 collected one from the "perpendicular volcanic rocks of New South Shetland" at latitude 62°S. It was *Usnea aurantiaco-atra*, one of the beard lichens.

Apart from sealers and whalers, nobody much—and especially not botanists—spent time in the Antarctic territory until 1839–1843, when James Clark Ross led a scientific expedition. J. D. Hooker, assistant surgeon and botanist for the expedition, noted, "The ships were already in the pack-ice, through which we penetrated, tracing the land to 64° and seeing a small volcanic island (Cockburn's Island), lying a few miles off the coast, we landed upon it. The vegetable productions only amounted to twenty Cryptogamic species, three of them Seaweeds." They turned out to be five mosses, ten lichens, and six algae. Hooker called the place a "desolate spot of land on the extreme limit of southern vegetation."

Antarctic exploration picked up toward the end of the nineteenth century when many nations (Belgium, Scotland, Great Britain, France, Germany, the United States) began sending expeditions. In 1895, Carsten Borchgrevink, aka Borchy, an Anglo-Norwegian, collected a lichen on one of the Possession Islands (around 71.5°S) while on a whaling expedition. "I quite unexpectedly found vegetation on the rocks, about thirty feet above the sea level, and considered this cellular and cryptogamous plant a lichen." Borchy collected more lichens at Cape Adare.

The most notable of these expeditions, from a lichen point of view, were the Belgian expedition of 1897–1899, which netted 55 species from the west coast of the Graham Land peninsula, and the second French expedition of 1908–1910, which yielded 112 species, of which 89 were new to science. Another boost came from Admiral Richard Byrd's second expedition in 1934 when the botanist Dr. Paul Siple led a small party in northwest Marie Byrd Land. For three months they collected lichens (and mosses) on the nunataks "by searching diligently on all exposures of rocks and in likely crevices wherever the party stopped." From their collections on 12 mountains, they found 89 species of lichen and five mosses.

Another significant contribution came from Elke Mackenzie (aka I. M. Lamb), who collected 1,030 specimens, 865 of which were lichens, during World War II as part of Operation Tabarin. "There were lichens in abundance: encrustments by penguin nests, sea cliffs, erratic boulders, shores studded by fossilized ammonites."

Lichen study in the Antarctic territory has become specialized, and the habitat is becoming more disturbed with thousands of visitors per year. That lichens live there at all is evidence of their ability to survive extreme weather. They make the primary contribution to the carbon cycle on the continent and could become important canaries in the climate-change coal mine.

See also Mackenzie.

Aposematism

Aposematism is a survival mechanism used by animals such as lichen moths to warn potential predators that they are either toxic or distasteful. The word comes from the Greek prefix *apo*, meaning "away," and *sema*, meaning "sign"—that is, a sign that you should keep away. It is pronounced ap-uh-sem-uh-tism. The warning is usually conveyed in the form of striking, stand-out coloration.

The common names of lichen moths that use this strategy are based on the bold coloring of the adult moth, for example, black and yellow lichen moth, painted lichen moth, and scarlet-winged lichen moth. They are in the same subfamily, *Arctiinae*, as tiger moths and are as boldly marked as many tiger moths.

Painted lichen moth
Hypoprepia fucosa

This is tangential, but it would not be right to mention the scarlet-winged lichen moth or its close relative, the painted lichen moth, without noting their well-developed skill in "fecal flicking." The caterpillars (okay, they're juveniles) can send their frass flying a distance of up to thirty body lengths. I don't know if they compete, but their fecal flicking does make it harder for parasites using scent to locate them.

On the spectrum of survival strategies, aposematism is at the opposite end from camouflage. Lichen moths are shouting, *Look at me! I taste dreadful! Eat something else!* Creatures using camouflage are quietly thinking, *We taste great but good luck finding us.*

See also camouflage; lichen moths.

Apothecium (*plural* Apothecia)

An apothecium is one of the two main types of spore-producing organs resulting from sexual reproduction of the fungal partner in most lichens. The term was first used in 1805 by Erik Acharius, known as the father of lichenology. The other type is a perithecium. (Asexual spores are produced in yet another structure called a pycnidium).

Most apothecia are shaped like discs, cups, or buttons. They may be attached to the top of the thallus or held above it on a stalk called a podetium. Don't expect to see them on all lichens—some lichens are sterile, some rarely produce apothecia, and some haven't gotten around to it yet (come back in a few years). On the other hand, some lichen species produce so many that what you see are wall-to-wall apothecia.

Apothecia are well-organized, complex structures. They contain sacs, called asci (singular ascus), within which the spores are formed. The asci are packed side by side, supporting each other vertically with some help from other specialized but infertile columnar fungal tissue (paraphyses). This dense crowd of sacs and columns is contained by a ring of yet another type of fungal tissue (exciple) holding it all together. Some species of lichen—the belt and suspenders group—go a step further and use some of their thallus to form an additional outer rim made of both fungal and algal cells.

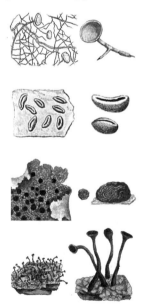

The open face of the apothecium can be red, blue, yellow, orange, white, pink, all the shades of brown, or black. The rim of the disc may be thick or thin; it may be colorless, the same color as the fertile center, or another color; and it may be ornamented with bumps or fibrils. Most apothecia are readily visible to the naked eye, although some are less than a milli-meter in diameter. A hand lens is helpful. The internal structure requires microscopy.

The word "apothecium" comes from the Greek *apothēkē*, meaning "storehouse" or "repository" (refer-ring to the storage of spores), albeit one with low inven-tory. In contrast to non-lichenized fungi, which need to disperse millions of spores in the short life of their fruit-ing bodies (for example, mushrooms), lichens have years to make and disperse spores. These sloths of the fungal world live at a different pace. When searching (under the microscope) for mature spores in an apothecium, you may find asci with immature spores or you may find the storehouse empty. It is time to behave like a lichen and persevere.

Not all lichens have apothecia, which is too bad because they add visual interest with their colors and shapes and provide an enormous amount of informa-tion about the identity of the lichen. Think of a lichen without apothecia as akin to a plant without flowers.

See also ascomycota; lips; perithecium.

Ascomycota

This phylum (pronounced ass-co-my-co-ta) is the larg-est in the kingdom Fungi with over 65,000 described species (and multitudes yet to be described). The name is derived from the Greek *askos*, meaning "sac," and *mukētes*, meaning "fungi." No surprise then that As-comycota are known as "sac fungi." All species in the phylum form their sexual spores in a sac called an ascus (plural asci), often inside a fruiting body. The species you are likely to be familiar with are the tasty and highly

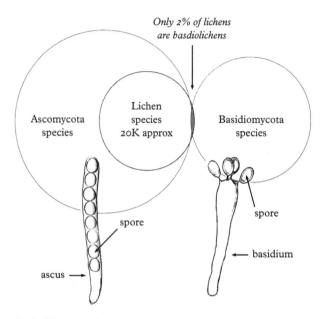

Only 2% of lichens
are basdiolichens

Ascomycota
species

Lichen
species
20K approx

Basidiomycota
species

spore

ascus →

spore

← basidium

desirable morels and the yeasts you use to make bread
or to brew beer. About two-thirds of all sac fungi, including morels, never form lichens. They have other
ways to get their food. On the other hand, almost all
lichens (98 percent) are formed from sac fungi. The
remaining 2 percent, called basidiolichens, are formed
from fungi in the phylum Basidiomycota, which have a
different mechanism for spore production.

The type of fruiting body where the asci are located
is a useful characteristic in narrowing down lichen identification. If the asci are located in a wide-open cup,
disc, or button on or above the lichen thallus, the fruiting body is called an apothecium. If the asci are deep

within a flask-shaped pit with only a pore opening on the surface of the thallus, the fruiting body is called a perithecium. Apothecia and perithecia are useful for getting to know lichens because you can see them without a microscope—and they are necessary names to know if you want to go deeper into lichen books.

The asci themselves, when seen at high magnification with a compound microscope, provide information that is often necessary for identification, especially for crustose species. Of particular interest are the thickness of the ascus wall, the tip where the spores are eventually released, and the number of spores per ascus, not to mention the many variations in the size and shape of the spores themselves. If you become captivated by lichens (and it's hard not to), you will want a compound microscope. The rewards will magnify.

See also apothecium; basidiolichens; perithecium.

Barnacles

Barnacles are marine crustaceans that build a protective shield of calcium plates around their bodies. Of the roughly 1,400 species, some are the bane of boaters. You may be familiar with the ones called acorn barnacles, which can be seen on rocks in the intertidal zone on North Atlantic and Pacific shores. They have two reasons to turn up in a lichen book. First, the calcium plates of acorn barnacles are a substrate for lichens. (Any surface that doesn't move around is sure to be colonized by a lichen.) Barnacles, after a few days as free-swimming larvae, choose a place to build their home of plates and never move again. Second, there are bark-dwelling lichens with apothecia that look so much

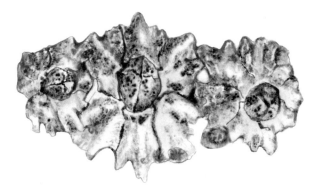

like barnacles that they are commonly called barnacle lichens.

A tiny marine lichen in the genus *Collemopsidum* lives embedded in a barnacle's protective plates. You can see the fruiting bodies (perithecia), looking like black specks smaller than one millimeter wide, on the surface of the plates. It also grows on other calcium-rich substrates in marine habitats, like limestone, limpets, and mollusks. The pitted surface of some barnacle plates is caused by the action of the lichen.

Another intertidal lichen that grows on limpets, *Thelidium litorale*, erodes the outer layer of the limpet shell in a curious way that makes the limpet resemble a barnacle. This lichen-induced mimicry is both good news and bad for the limpet. Its shell is now weaker, but on the plus side, it is less likely to be eaten by shorebirds who seek out limpets over barnacles.

Barnacle lichens that grow on tree bark are in the genus *Thelotrema*. *Thelo* is from the Greek for "nipple," and *trema* means "hole" or "aperture." The apothecia somewhat resemble nipples with a hole at the top. Think of a volcano with flattened sides, or better, think of a barnacle. One species in particular, *Thelotrema lepadinum*, also known as the tree barnacle, has apothecia that look exactly like miniature barnacles. The lichen thallus is a thin pale crust, and its apothecia are usually abundant, just like barnacles on a rock.

See also apothecium; perithecium.

Basidiolichens

Basidiolichens make up a small percentage of lichens with a mycobiont (fungal partner) from the phylum Basidiomycota, within the kingdom Fungi. This large phylum contains most of the fungi that produce familiar edible mushrooms such as portobellos, chanterelles, oyster mushrooms, and puffballs, none of which form lichens. The phylum is so named because its members form their sexual spores on the end of a club called a basidium. (The Latin word *basidium* means "little pedestal." Most mycologists would agree that a word meaning "little club" would have been a better choice.)

Only about 2 percent of lichens are formed from these types of fungi. The remaining 98 percent use fungi from another phylum, Ascomycota, in which the spores are formed inside a sac. These sacs and clubs are microscopic and very important to taxonomists. You need to mount a section of a fruiting body on a slide and look at it with high magnification on a compound microscope. You will never see these sacs and clubs in the field. Come to think of it, you probably won't see taxonomists in the field either—they are busy in the lab with their DNA sequencing.

*Lichenomphalia
umbellifera*

I'm singling out basidiolichens here because they don't look like other lichens. The fruiting body is the most conspicuous part. It looks like a mushroom—because it actually is a mushroom. For this reason, basidiolichens in the genus *Lichenomphalia* have the common name mushroom lichens.

That was the simple story of basidiolichens—everything very neat and nice—before we knew that the predominant fungus is not the only fungus present. In addition to numerous other lichenicolous fungi, Toby Spribille discovered that another fungus, a yeast, is often present, and it happens to be in the phylum Basidiomycota. So, in that sense, many lichens are partly basidio.

See also ascomycota; lichenicolous fungi; quandaries.

BIOMEX

The European Space Agency (ESA) and the Russian space agency Roscosmos conducted the Biology and Mars Experiment (BIOMEX) between 2014 and 2016 in order to investigate the survivability of life forms and their potential for inhabiting Mars. Lichens are known to survive extremely harsh environments on Earth, but how would they do on Mars?

In earlier ESA experiments, two rock-dwelling lichens (*Rhizocarpon geographicum*, yellow map lichen, and *Rusavskia elegans*, elegant sunburst lichen) were exposed to ten days of the extreme temperatures, cosmic radiation, and vacuum of space. After the return to Earth, results showed that fungal spores of both species were still able to germinate and grow. The growth rate of the algal cells of the *Rusavskia* did suffer serious slowdown—but not death!

Other experiments tested the limits of survival and delved into lithopanspermia—the idea that life could have been transferred through space, between solar systems, on rocks and may have arrived on Earth from meteorites. In the end, two lichens, *Circinaria gyrosa* and *Buellia frigida*, were chosen for BIOMEX—along with archaea, bacteria, cyanobacteria, an alga, a moss, a liverwort, a black endolithic fungus found in Antarctica, a kombucha biofilm containing yeasts and bacteria, eight types of biomolecules (including chlorophyll, cellulose, and chitin), and an assortment of minerals and substrates. The samples were fully exposed to the conditions of space on the exterior of the International Space Station from October 22, 2014, to February 3, 2016.

The ESA website reports that "lichen and small organisms called tardigrades or 'water bears' have spent months outside the International Space Station and have returned to Earth alive and well." Scientists on the BIOMEX experiment reported that the lichens were more susceptible to the exposure than the single-cell organisms and had suffered, in their words, "a significant decrease in vitality," to which I can only say, no kidding.

See also tardigrades.

Black Stone Flower

This poetic common name belongs to a lichen—not a flower—also known as the powdered ruffle lichen, and in India as Kalpasi and Dagad Phool. The scientific name is *Parmotrema perlatum*.

It is a foliose lichen, mineral gray on the upper surface with a black and brown lower surface. It is found on all continents, usually growing on trees and sometimes on rocks. In India it is collected, dried, sorted, and packaged for sale as a spice. It is a necessary ingredient in goda masala, a spice blend used in Indian cuisine, especially Maharashtrian dishes. Garam masala, used in Punjabi and northern Indian recipes, could be substituted, though it has more heat and pungency than goda masala.

Black stone flower is inert in its dried state but releases an aroma when heated with a little oil to make the masala, to which it lends an earthy flavor and a deeper color. Because lichens tend to be bitter to the human palate, they are boiled in water to remove (or reduce) the bitterness before being used in food. When used in masala, however, the bitterness may be desirable and needs to be preserved.

Many different lichens are collected in India for use in spice blends and in the perfume industry, as well as for medicinals. (In the Ayurvedic tradition of medicine, black stone flower is suggested for a wide variety of ailments.) Concerns about sustainability have raised the possibility of limiting the trade of some species and adding them to the Convention of International Trade in Endangered Species of Wild Fauna and Flora (CITES) list.

See also food (for humans).

Black Tree-Hair

Black tree-hair (*Bryoria fremontii*) is a filamentous lichen that grows in fine tangled strands up to two feet long (there's a reason it's called tree hair) on species of pine, fir, and western larch. Other common names are tree-hair lichen and, in the absence of marketing people, edible horsehair. It is one of the few lichens with a long history of use as food for humans. Most accounts are from tribes in western North America. Some ate it only during famine, while others ate it by choice, possibly because the taste varied according to the tree it grew on, the method of preparation, or the other lichens inadvertently collected with it.

It was important to distinguish black tree-hair lichen from yellow horse-hair, *Bryoria tortuosa*, which shares the range, also grows on pines, and is often visually similar. A toxic chemical, vulpinic acid, is present in both species, but in high enough concentrations in yellow horse-hair to make it one of the few poisonous lichens. Collections for food were tested by taste to judge the degree of bitterness.

A customary method of preparation was to cook black tree-hair slowly for up to two days in a pit on a bed of hot rocks covered with mosses, ferns, grasses, and skunk cabbage leaves to generate steam. Sometimes it was cooked on its own, and sometimes it was mixed with wild onions, camas bulbs, Saskatoon berries, or other flavorings. The lichen cooked down into a black gooey paste; some was eaten right away, and some was dried into "cakes" that would keep for as long as three years. Another method was to mix black tree-hair with camas, dry it, and powder it, creating a luxury food. Yet another method was to pile it up, leave it to ferment, then roll it into balls as big as your head and bake them in a pit oven.

There is a story from British Columbia of a man weighting down a clump of *Bryoria* in a lake overnight. Soaking it in a lake—or better, in a stream—was a standard way to clean it and get rid of twigs and other forest debris. He didn't expect to find it covered with freshwater shrimp when he retrieved it the next day but took advantage and made lichen and shrimp soup. He liked it enough to repeat it.

Black tree-hair's nutritional value is based on its relatively high percentage of carbohydrate along with a small quantity of protein, but it was also believed to have medicinal properties. Okanagan people mixed dried, powdered tree-hair with grease and rubbed the resulting paste on newborn babies' navels to prevent infection. Nez Perce people thought it useful for upset stomach, indigestion, and diarrhea. Scientists today are still trying to understand and uncover the potential of lichen substances for a variety of medicinal uses.

The fibrous nature of the lichen suggested yet another example of making the most of what you have. Some groups fashioned it into clothing. Long strands, pliable when wet, were cleaned, twisted, laid out in rows, and woven into shapes with strips of wolf-willow bark. Tree-hair lichen garments (moccasins, vests, and leggings) may have been okay in dry weather. As indoor slippers, lichen moccasins might provide insulation, and if skins were scarce, they would be better than nothing for outdoor wear. Another application of the lichen's insulation potential can be seen in an account of a Lillooet family that chinked the cracks in their log cabin with bags of raw black tree-hair lichen.

See also coyote; food (for humans); quandaries.

British Soldier

The name of this lichen comes from the red coats worn by the British soldiers in the American Revolutionary War. The lichens could be more accurately described as having red caps rather than red coats. The purpose of the soldiers' garb, however, was to make the troops stand out and be noticed, so in that sense the name is perfect. You can spot these lichens from a distance because of the color.

The term "redcoat" was originally used to describe British soldiers in Ireland during the first Elizabethan era (1558–1603). The Irish continued to use the term through many other battles with the British, and it is thought that Irish immigrants brought "redcoat" to America.

The British soldier lichen is native to North America and occurs in the eastern half of the continent from

Cladonia cristatella

arctic latitudes down to Florida. It was first described in 1858 by Edward Tuckerman, who assigned it to the genus *Cladonia*, with the epithet *cristatella*. The name stands today, despite the fact that Tuckerman was accused of having "no clear conception of genera."

British soldier lichens often grow in troops and are quite common—you can find whole armies of them camped out in sunny spots on soil and on dead wood like fence railings or wooden benches. They start out low on the substrate as little green fragments called squamules. At that stage, it's not possible to identify them. As they mature, they develop many grayish-green hollow stalks up to an inch tall called podetia, which are sometimes branched and always terminate in bright red smooth and rounded caps.

This lichen has the distinction of being the first lichen to be reliably resynthesized in a lab after its components had been teased apart. Many had tried, but it was Vernon Ahmadjian, an expert in lichen biology at Clark University in Worcester, Massachusetts, who succeeded in the 1960s.

The one lichen that people in northeastern North America are likely to know is British soldiers, but they can be fooled. The actual British soldiers marching to the Old North Bridge in Concord, Massachusetts, for the skirmish that would launch the revolution were unmistakable. Alas, their namesake lichens are not so distinctive. They have relatives with strong family resemblances, all wearing red caps. A close inspection is needed to distinguish true British soldiers (*Cladonia cristatella*) from gritty British soldiers (*Cladonia floerkeana* if you're formal, or gritty Brits if you get familiar), powderfoot British soldiers (*Cladonia incrassata*), southern soldiers (*Cladonia didyma*), and toy soldiers (*Cladonia bellidiflora*). My favorite family member didn't join the military. It has a tall, slender, unbranched podetium and a modest red cap at the tip. It is called lipstick lichen (*Cladonia macilenta*). If you confuse a British soldier with a lipstick, it's time for cataract surgery.

See also Tuckerman.

Brodo, Irwin (1935–)

Now and then in every field of endeavor, a game changer turns up. For people interested in lichens, this is Irwin

"Ernie" Brodo, Professor Emeritus, Research and Collections Division, Canadian Museum of Nature. He was recognized in 1994 for his many contributions to the field with the award of the Acharius Medal, the highest honor given in lichenology. A giant contribution was yet to come.

The breakthrough event occurred in 2001 with the publication of his *Lichens of North America*, written with Sylvia Duran Sharnoff and Stephen Sharnoff. In just under 800 pages of a large-format book, amateur naturalists can learn everything they might want to know about lichen lifestyle and nomenclature, plus keys to identifying the species and detailed descriptions of 800 species with photos, notes, and comments about an additional 700 species. The color photographs by the Sharnoffs make the book into a masterpiece of art. The images are so much more than documentation of nature—they make lichens strut their stuff as superstars on nature's runway.

The book is clearly not a field guide, unless you take a wheelbarrow with you, but that's not how it was intended to be used. *Lichens of North America* was the first singular and comprehensive resource that anyone with an interest in lichens anywhere in North America could use to further that interest. It was followed in 2016 by a supporting book with revised and expanded keys, covering 2,000 species. Serious amateurs were equipped to build their own knowledge and to add to the scientific body of knowledge. They also had the material to host classes for schools and nature groups. Brodo and the Sharnoffs had brought lichens out of obscurity.

Camouflage

Camouflage
Many animals use lichens for camouflage, taking advantage of their presence in almost every kind of habitat, from arctic to desert. Lichens have been around for millions of years—enough time for a variety of camouflage mechanisms to develop and become effective for different species on different continents.

Some animals' entire bodies resemble lichens in both color and form. For example, the lichen katydid, *Markia hystrix*, which lives among lichens (often *Usnea* species) in Central and South America, has a lichen-green body with numerous spiky projections exactly mimicking both the color and the form of the fruticose lichen it lives with. Similarly, the common bark katydid of Africa, *Cymatomera denticollis*, has lobed growths on its legs and thorax that resemble the shape and color of the foliose lichens in its habitat.

Many animals rely only on color for camouflage. One of the largest such animals is the Colugo, whose blotchy body blends with lichen patches on tree trunks. This

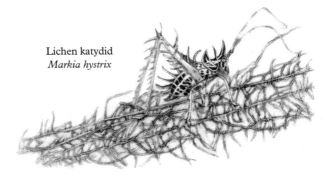

Lichen katydid
Markia hystrix

background matching is very common among *Lepidoptera* (mainly moths) larvae, adults, or both. Larvae that feed on lichens and nocturnal adults that hang out during the day on lichen-covered bark—such as the European tree-lichen beauty *Cryphia algae* and the gray-green Brussels lace moth *Cleora lichenaria*—use this method. In tropical rain forests, white lichens grow directly on living leaves, giving them a speckled appearance. The Costa Rican katydid *Rosophyllum colosseum* is green like the leaf and has the same whitish specks, allowing it to blend perfectly with the leaf. The tropical tree frog *Hyla rufitela* uses the same strategy.

Some spiders are adept at lichen camouflage. In the Wassaw Wildlife Refuge off the coast of Savannah, Georgia, I once found a little green spider (*Eustala sp.*) on the medallion lichen (*Dirinaria confluens*). Not only was the green a perfect match with the lichen, but the center of the spider's abdomen was black—exactly the color and size of the apothecia on the lichen. If it hadn't moved, I would not have noticed it.

Some moths hedge their bets by having both dark bark-mimic forms and pale or greenish lichen-mimic forms. Examples are the adult of the peppered moth, the caterpillar of Kent's geometer, and the caterpillar of the lunate zale.

Other creatures using lichen camouflage have to work to achieve the same result as creatures that are born disguised. Caterpillars of bagworm moths create cases of bits of plant debris and lichen—whatever is convenient. They retreat into these camouflaged cases when threatened or when resting. Some species of lacewing larvae, known as trash carriers, accumulate bits of wood, bark, remains of insects and spiders, and other debris into a pile that they carry on their backs. One lacewing species, *Leucochrysa pavida*, is a specialist: this lacewing collects only bits of lichen. Its lichen "coat" may be four times wider than its own body, making the coat look like a patch of lichen.

A singular example of lichen actually growing on an invertebrate is found in the New Guinea forest weevil (*Gymnopholus lichenifer*). When they first emerge as adults, these weevils are up to one and a half inches long, very slow moving, don't fly, and are free of lichens. A tiny microhabitat consisting of mites, nematodes, rotifers, bark lice, and lichens builds up on their backs. Some of those critters may bring in lichen fragments that later get established. It is thought that orabatid mites, which are voracious lichen eaters, keep the lichen in check.

See also colugo; hummingbird nest; peppered moth.

Careers

Lichenology has emerged from obscurity, shaken off a somewhat dowdy image, and become an enticing field for young scientists. The new breed of lichenologists working today is typified by James Lendemer, Staff Lichenologist and Associate Curator of the Institute of Systematic Botany, New York Botanical Garden (NYBG). While no less disciplined in their science than their predecessors, these younger lichenologists are bringing lichen texts out of the academics-only arena to a place of general interest. In so doing, they engage budding naturalists and journalists in the popular media, who bring lichen stories to the general public where everyone can feel the wonder.

Lendemer's interest in lichens began when he was a teen helping with collections at the Academy of Natural Sciences in Philadelphia. He experienced the institution's stated mission of "encouragement and cultivation of the sciences" when he discovered a mistake in the naming of a lichen specimen. Despite his lack of status, lichenologists gave him respect for his observation. This experience led to other discoveries: lichens are pretty amazing, lichenology is in need of more attention, herbaria collections are important, there's plenty of room for newcomers, and there's support among peers. Lendemer was hooked, and he's now paying it forward. He's a lichen evangelist, in the best sense of the word.

In addition to the awesome responsibility of overseeing NYBG's lichen collection (the largest in the Western Hemisphere), Lendemer works on documenting the lichen diversity of North America and the importance of lichens in conservation strategies for land managers. His collaboration with other lichenologists, environmentalists, conservationists, and rock climbers has led to the discovery of new species, expanded the range of known species, encouraged habitat protection for threatened species, heightened awareness of endangered species, and produced user-friendly field guides.

The power of collaboration makes two recent field guides shine. *Urban Lichens*, Lendemer's collaboration with lead author Jessica Allen, Associate Professor of Biology at Eastern Washington University, not only opens readers' eyes to the fact of lichens in urban habitats but also details how to recognize them. It even includes an itinerary for "Fifty Lichens in One Day: A Whirlwind Tour of New York City."

The other is *Field Guide to the Lichens of Great Smoky Mountains National Park*, a collaborative effort between Lendemer and lead author Erin A. Tripp, Associate Professor, Ecology and Evolutionary Biology, and Curator of Botany, University of Colorado. The guide is more than user-friendly—it's lichen-friendly. The authors unabashedly express their personal feelings about lichens while still including the necessary formal details. "*Punctelia* is a wonderful genus of macrolichens whose species are quite at home in the Smokies, like many of the rest of us." The range maps show that this "wonderful genus" might also be at home where you live. They call one of the weediest species in the genus, *Punctelia rudecta*, not a weed but "Backyard Buddy." Their joy in lichens is infectious. These young lichenologists and their peers are spreading the love.

See also Parton, Dolly; rock climbing; Wyoming.

Colugo

This curious animal, the "flying lemur" of Southeast Asia, neither flies nor is it a lemur. If it weren't for its excellent daytime camouflage, it might have been given a more accurate name, like "big-eyed, giant lichen lurker." Many animals lurk on and in lichens, but most are very small, and many are invertebrates. The colugo is a furry mammal somewhere in size between a gray squirrel and a household cat. It is a canopy-dwelling nocturnal animal that spends its days in tree hollows or clinging to tree limbs with its long, curved claws. Some individuals have reddish fur, and others have mottled gray or greenish-gray fur with blotches of a paler color that allow them to blend in with lichen-covered tree trunks and escape detection from predators, which include pythons, martens, macaques, and owls.

In Central and South America, the sloth is another arboreal mammal with greenish fur, owing to the green algae that live in its fur. In fact, there is a three-way dependency among the sloth, the algae, and a moth that also lives in the sloth's fur. It is a less intimate algal symbiosis than that of lichens. The algae seem to mainly function as a food supplement for the sloth and a specialized mini-habitat that supports the entire life cycle of the moth. Another function may be providing camouflage for the sloth.

Just how strange a mammal is the colugo? It is closer in its evolutionary history to primates than to bats (true flyers) or flying squirrels (furry gliders). Taxonomists have found no evidence to align it with any other mammals and have assigned it an order of its own, the Dermoptera, a name that means "skin wings." A skin membrane called a patagium connects the sides of the head, the front leg claws, and the hind leg claws; it runs all the way to the tip of the tail, giving the animal maximum surface area to support efficient gliding.

Whereas other mammalian orders include hundreds of species—or in the case of rodents over 2,000 species—the Dermoptera has only two for sure and maybe no more than eight. As members of the Primates, we may think of ourselves as distinctly different from the other 500 primate species, yet we have enough in common for the taxonomists to have lumped us all together. But there is nothing like a colugo except another colugo.

See also camouflage.

Common Names

I liken (pardon me, but it's unavoidable now and then) common names to nicknames. They indicate affection and familiarity, but in biology they are problematic because they are not uniformly agreed upon and differ across languages, regions, and cultures. Scientific names, on the other hand, are backed up with precise descriptions and worldwide acceptance but have a major shortcoming. They use Latin, which is not widely known and is alienating. In contrast, common names are often endearing, amusing, fanciful, and in any event, engaging.

Common names are extra important for lichens because they help interest people in a part of the natural world they have overlooked. In the northeast United States, birthplace of the American Revolution, many people who don't know lichens are aware of a red-capped one called British soldiers, but they have never heard of *Cladonia cristatella*.

Most lichens are diminutive with distinctive structures that have led to fanciful names like pixie cups, gnome fingers, elf ears, angel's hair, and witch's hair. The fairy puke lichen, with its blue-green thallus splattered with pink blobs, surely commemorates what happened when the fairy heard that its scientific name is *Icmadophila ericetorum*.

Yellow lichens in the genus *Vulpicida* are called sunshine lichens, and the bright orange lichens of the *Xanthoria* genus are known as sunburst lichens. Common names for individual species narrow down the description, but they're subjective. Some people may find the bare-bottomed sunburst just as elegant as the elegant sunburst.

Some common names are beautifully simple. The asterisk lichen looks exactly like a group of asterisks drawn with a black pen on the bark. The smoky-eye boulder lichen grows on boulders and each little gray "eye" looks ringed with black eyeliner. The Christmas lichen is red and green. When I have shown a collection of toadskin lichen, *Umbilicaria papulosa*, to a class of fifth-graders and asked what they would call it if they had to assign a name, someone always says, "It looks like a toad."

Animal names are popular. Reindeer lichens are well known, but there are also dog, wolf, owl, mouse, and centipede lichens. There are foxhair lichens and horsehair lichens. A favorite is the can-of-worms lichen.

Mealy pixie cup
Cladonia chlorophaea

My birding friends refer to birds by their common name. My lichen friends call lichens by their scientific names. Would we have broader lichen interest and awareness if we used common names? To say nothing of greater fondness for them? After all, lichens don't poop on your car.

See also British soldier.

Coyote

Coyote features in many Native American legends, sometimes as the primary character and sometimes with other animals like Fox and Eagle. He can be foolish but is also resourceful and wily. Coyote is a compulsive trickster, a troublemaker, and, very importantly, a transformer.

A legend of the Okanangan-Colville people of inland portions of Washington State and British Columbia tells of the origin of the black tree-hair lichen (*Bryoria fremontii*). As happens in a culture of oral traditions, there are multiple variations of the legend. One version tells of Coyote and his son capturing some swans. He ties his son to the birds to hold them while he climbs a nearby pine to get kindling wood. When he is way up the tree, the swans fly off with his son. He tries to jump out of the tree, but his long hair braid catches on a branch and he is stuck. He eventually frees himself by cutting off his braid. Meanwhile, the swans have dropped his son, killing him. Coyote, the transformer, not only revives his son but also converts his hair into a long black stringy form of *Bryoria* lichen. He addresses it, "You, my valuable hair, will not be wasted. The people will gather you and the old women will turn you into food."

This lichen is one of the few that to this day is eaten out of choice rather than only in times of famine.

See also black tree-hair.

Crustose

One of the three major growth forms of lichens is called crustose. These lichens are bonded so closely to their substrate that they can't be collected without taking some of the substrate with them. Like foliose lichens, they have an upper cortex, but they differ in having no lower cortex. On the underside, microscopic fungal threads extend directly into crevices in the substrate and hold the lichen in place. Crustose lichens exist like a skin on the substrate. Some are thin and almost two-dimensional; others are thick and lumpy. Textures range from baby's bottom smooth to elephant trunk bumpy. Similarly, the surface may be smooth and continuous, coarse, lightly cracked, or deeply cracked.

Two types of crustose lichens stretch the definition of the group. One type is endolithic, meaning it grows within rock. The fungal threads and algal cells find enough space between the grains in the first few millimeters of the rock to survive and grow. Only the fruiting bodies (apothecia) are on the rock's surface. The other type is called leprose because the thallus is broken up into powdery fragments that create a rough, scurfy texture, typified by the genus *Lepraria*, known as dust lichens.

Crustose lichens can form colonies that cover large surface areas and are visible from a distance, especially brightly colored ones, like yellow map lichen, *Rhizocarpon geographicum*, on otherwise bald granite outcroppings; orange desert firedot, *Xanthomendoza trachyphylla*,

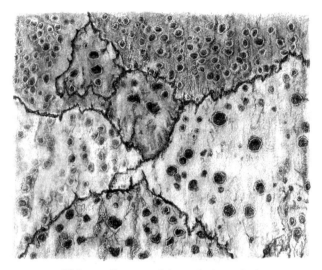

Wall-to-wall crustose lichen colonies on bark

on exposed rocks in very dry habitats; and maritime sunburst lichen, *Xanthoria parietina*, on coastal rocks.

Lichens of different species and different growth types are often found in very close proximity, growing into and over each other while they cover large rocky areas so thoroughly that the natural color of the rock is hidden. Crustose lichens on bark can completely obliterate the color and texture of the bark, interfering with identification of trees by bark. Erect (fruticose) types tower over crustose types on the same twig. Another feature of some crustose species, which is most apparent when colonies abut, is a black border outlining the boundary of each colony, like lines delineating countries on a map or patches on a crazy quilt. This border material, called a prothallus, is purely fungal and often black. You may also notice it in the fissures of crustose lichens with a cracked surface.

See also foliose; fruticose; thallus.

Cudbear

The purple dye industry started as early as 1500 BCE in southern Europe, with Mediterranean shellfish as the source, and later moved on to *Roccella* lichens, the source of the dye called orchil. The dates are hazy, but we know that purple dyes were being extracted from other lichens in northern countries at roughly the same time. Much later, some of these dyes came to be known as cudbear.

Jump ahead to Scotland and the time of the Industrial Revolution. The Highland Clearances were driving tenants off the land to make way for higher-rent sheep farms for wool for the mills. Meantime, the Gordon family of Banffshire had figured out a process for making purple dye from Scottish lichens. Their patent application of 1758 read in part, "When these three Ingredients [lichens] are gathered, cleanse them from all their filth, by laying them severally in cold water, changing the water daily so long as any filth remains about them. Then dry and pound them in a mortar and dilute them with the spirit of urine & spirit of soot, to which add quick lime. Digest them together for fourteen days, and they will produce the Cudbear fitt for Dyers use." The Gordons called their dye "cudbear" for Cuthbert, which was Mrs. Gordon's family name. A benefit of cudbear over imported *Roccella* from Cape Verde and the Canary Islands was that it was entirely available from "Great Britain or His Majesty's plantations." They could set their own price.

While the Gordons prospered with cudbear factories in Edinburgh and Glasgow, the displaced Highlanders struggled. They collected the lichen *Ochrolechia tartarea* in the largest quantities they could find, and they saved their urine for sale. At one point, 2,000 to 3,000 gallons per day were needed. Collectors used hygrometers to prevent fraud. As Scottish sources of the lichen were depleted, imports from Norway and Sweden filled the gap. Eventually, the enterprise failed.

Cudbear manufacturing in England also sourced lichens from Norway and Sweden. The process was basically the same: it still involved ammonia and the common source was still urine. William Lauder Lindsay (1829–1880), a Scottish physician and botanist specializing in lichens, reported that English manufacturers had "learned by experience to avoid urine from beer-drinkers, which is excessive in quantity but frequently deficient in urea and solids, while it is abundant in water."

Lichen flora got a chance to recover starting in 1856 when a London teenager, William Henry Perkins, accidentally synthesized a purple dye while trying to make quinine for treating malaria. All dyes up to this time had come from natural sources. Perkins understood the significance of his discovery, started a dye business, and subsequently discovered other dyes.

See also dyes; orchil.

Cyanobacteria

About 10 percent of all lichen species contain cyanobacteria, either as the sole photobiont or as a supplement to algae. Cyanolichens are relatively young in the tree of life, but cyanobacteria are older than the hills. They have been around for three and a half billion years. Some of them floating near the surface of the ocean stumbled upon the process of trapping solar energy and using it to construct organic compounds, giving off oxygen. Hello, photosynthesis! Among the compounds they made was the enzyme nitrogenase, which is used for fixing nitrogen. Over time other primitive life forms developed and engulfed cyanobacteria, presumably for food. The accepted theory is that some engulfed cyanobacteria continued to function inside the host. Eventually, they gave up free-living and nitrogen-fixing, kept on photosynthesizing, and evolved into chloroplasts.

Cyanobacteria are still with us, and many are free-living; some species (frequently *Nostoc*) are captured by fungi and form cyanolichens, for example, *Peltigera* or *Collema*. A few species are edible by humans. For example, *Spirulina*, which is high in protein, vitamins, and minerals, was a food source for Aztecs. Today it is

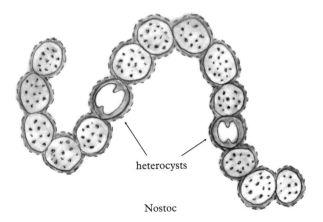

heterocysts

Nostoc

sold in the United States as a "health food" in powder or tablet form. NASA has used it as a dietary supplement for astronauts. Other edible examples are some *Nostoc* species that are used as food in the Philippines, Indonesia, Japan, and China. On the other hand, many species (in at least eleven genera) produce a variety of cyanotoxins that target the liver, skin, or nervous system.

Cyanobacteria used to be called "blue-green algae" because, like algae, they photosynthesize, but biochemistry and electron microscopy have shown them to be a form of bacteria. They contain chlorophyll *a*, and many also contain the blue pigment phycocyanin, which is also capable of photosynthesis and which gave them their name.

Cyanobacteria had a gigantic effect on the development of life on earth. They gave us chloroplasts and an atmosphere with oxygen. Sir Isaac Newton got a lot of things right, but he should have said, "We have come this far by standing on the shoulders of microbes."

See also cyanolichens; nitrogen fixing.

Cyanolichens

The majority of lichens, around 90 percent, get their carbohydrates from algae. The remaining 10 percent, the cyanolichens, get some or all of their carbs from cyanobacteria. You can recognize a lot of cyanolichens just by looking at them or, for some species, by smelling them. If you come across a gray, brown, or black lichen that smells like fish or that you want to call a jelly lichen, chances are good that you have found a cyanolichen. As you might expect with lichens, however, it's not always that simple. One of the curveballs is that lichens with both algae and cyanobacteria are probably more green than gray, brown, or black. These three-part lichens (one mycobiont and two photobionts) rely on algae as the primary photobiont and contain cyanobacteria in nodules or galls called cephalodia, which appear as dark or bluish lumps. The main role of cyanobacteria in three-part lichens is nitrogen fixation. The frequency of special cells for nitrogen fixation, so-called heterocysts, is much higher in the cyanobacteria in cephalodia than in two-part cyanolichens. Examples of three-part cyanolichens are some of the dog lichens (*Peltigera*), foam lichens (*Stereocaulon*), and lungworts (*Lobaria*).

In two-part cyanolichens (one mycobiont and one photobiont), the cyanobacteria are not confined in cephalodia. If the lichen is gelatinous when wet, you can probably narrow its identity down to one of the jelly lichens (*Collema*) or jellyskins (*Leptogium*), in which the cyanobacteria are spread throughout the thallus. If the cyanobacteria are confined to a layer, you might have a nongelatinous shingle lichen (*Pannaria*).

If a lichen is dark brown or gray brown, foliose, and smells fishy, you have probably found a moon lichen (*Sticta*). James Lendemer, Staff Lichenologist (and Lichen Sniffer) as well as Associate Curator of the Institute of Systematic Botany at the New York Botanical Garden, has offered the appropriate nickname of "Stink-ta" for *Sticta beauvoisii*, one of the moon lichens.

Caribou are famous for consuming reindeer lichens, which don't have cyanobacteria, but they do resort to eating certain cyanolichens (*Stereocaulon*) in the winter in parts of the boreal forest where cyanolichens are the dominant ground cover.

Although cyanolichens produce glucose to feed the fungus, and lichens with green algae produce sugar alcohols of a kind similar to those used as sweeteners in the food industry, lichens do not taste sweet. Any sweetness is masked by the many bitter substances produced by the fungus.

See also cyanobacteria.

Dermatology

The word "lichen" has been used to describe skin ailments for almost as long as the word has existed. Hippocrates (ca. 460–370 BCE) used the word to mean "eruptions of papules." Dioscorides (40–90 CE) used it to mean both the lichen *and* the "papular skin disease" it treated. Today the word is part of the official dermatology lexicon, which is no wonder given the visual similarity of some lichens to some skin ailments.

"Lichenification" is a general term for the secondary condition of hardened, thickened, cracked skin caused by chronic itching and scratching. The medical name is "lichen simplex chronicus." It is not restricted to humans; there are accounts of it on black bears, golden eagles, and bobcats. Crustose lichens described as areolate (for example, tile lichens in the genus *Lecideia*, or cobblestone lichens in the genus *Acarospora*) or rimose (such as map lichens in the genus *Rhizocarpon*) have a thick thallus and a cracked surface that resemble lichenification—at least in black and white.

Dermatologists use the term "lichenoid eruption" to describe several skin disorders that involve flat, shiny lesions called papules. "Lichen planus" is characterized by irritated areas with papules that could easily resemble a lichen thallus with apothecia. The discolored patches of "lichen schlerosus" contrast with surrounding healthy skin, making a pattern like a lichen on a rock or tree trunk. And so it goes with drug-induced lichenoid eruption, lichen nitidus, lichen scrofulosorum, lichen striatus, lichenoid pityriasis, and exudative discoid and lichenoid dermatitis!

Lichens have nothing to do with leprosy, but some species with no protective skin (upper cortex) have a powdery surface and are called leprose. They are typified by the *Lepraria* genus, aka dust lichens, and *Chrysothrix*, aka golddust lichens.

A few individuals are allergic to some species of lichens and can develop contact dermatitis. This is mostly seen in forestry workers.

If you don't want to experience a cringe-fest, skip the search for lichen-named skin diseases in image libraries.

See also etymology.

Dispersal

Lichens seem to face massive challenges to their dispersal. Only the fungal partner reproduces sexually, and in North America around 23 percent of lichenized fungi never do. Spores resulting from sex are produced in apothecia and created over the long life of the lichen. Some species produce huge numbers of apothecia (more spores, more dispersal opportunities); some ground-dwelling species, such as *Cladonia*, produce apothecia on a stalk, where released spores have a better chance to catch a current of air.

The "celibate" ones are creative about vegetative reproduction. They have a variety of specialized reproductive propagules, all designed for dispersal. The propagules are easily dislodged by heavy weather, falling branches, wind, water, and animals. Almost all lichens are brittle when dry, such that any disturbance might break off a piece regardless of special structures.

Whatever the method, lichens are successful. They occupy 6 to 8 percent of the Earth's surface and an unknown percentage of tree bark and other surfaces.

See also isidium; reproduction; soredium; zoochory.

Doctrine of Signatures

This medieval doctrine held that plants with useful medicinal properties had been marked by God such that their shape, color, smell, or texture indicated the body part or ailment that they could be used to treat. These beliefs were popularized and practiced at a time when lichens were still considered plants, so it's not surprising to find lichens in the doctrine.

Beard lichens (genus *Usnea*) are filamentous, and although green, their hairlike form suggested to these early observers that they could strengthen hair and treat conditions of the scalp. It turns out that beard lichens do produce an antibacterial compound, called usnic acid.

The maritime sunburst lichen, *Xanthoria parietina*, signaled its possible use for treating jaundice by its orange color. Many lichens are yellow or orange. Maybe this one was more prolific in an area with avid practitioners of the doctrine. None of the orange lichens are effective for jaundice.

The dog lichen, *Peltigera canina*, has apothecia that, with some imagination, look like dog teeth—albeit brown dog teeth. Another possible signature is the white

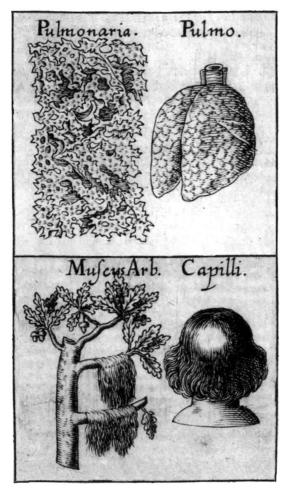

Depiction of lungwort and beard lichen by
Michael Bernhard Valentini, *Medicina nov-antiqua*, 1713

pointed rhizines on the lower surface that could resemble dog fangs. Either way, concoctions of dog lichen were recommended for the treatment of "canine madness" or rabies. John Lightfoot, an English botanist and parson who didn't subscribe to the doctrine of signatures, wrote in 1777: "It is much to be lamented that the success of this medicine has not always answered the expectation. There are instances where the application has not prevented hydrophobia; and it is even uncertain whether it has been at all instrumental in keeping off that disorder." Mr. Lightfoot was a master of understatement—his light touch matched his foot.

Lungwort, *Lobaria pulmonaria*, has a textured surface that is evocative of lung tissue and therefore indicated to believers in the doctrine a treatment for tuberculosis. Unlike the connection drawn between dog lichen and rabies, this association has some merit: lungwort contains antibacterial agents that do have an effect on tuberculosis.

Among the many compounds unique to lichens, some are truly medicinal.

Dyes

Lichens have been used as a source of dye since classical Greek times and since unknown times by indigenous peoples. Materials dyed have included wool, silk, cotton, skin, basket reeds, spruce roots for baskets, and porcupine quills. The colors achievable cover the whole spectrum from reds and yellows to purples and violets. There is no relationship between the lichen color and the dye color; purple dye, for example, can be extracted from the gray-green lichen *Parmotrema tinctorum*, and blue dye from the orange lichen *Xanthoria parietina*. The dye color is a result of chemical changes in lichen compounds during the dye process. The dye compounds are of two types, each of which has its corresponding extraction process.

The boiling-water method typically yields yellows, golds, and a range of browns; it depends on compounds like stictic, norstictic, or salazinic acid. Dye lichens in the *Parmelia* genus (called "crottle" in Scotland and Ireland) contain salazinic acid, which converts to a dye compound that forms such a stable bond with wool proteins that no mordant is needed. The earthy tones and earthy aroma (some call it musty) of the original Harris Tweed came from crottle. Wool dyed this way had the added value of being too bitter for moth caterpillars.

The ammonia method involves steeping the lichen in the presence of ammonia for three to sixteen weeks, depending on the lichen species. The old-time source of ammonia, urine, was later replaced by ammonium hydroxide. This method depends on compounds like erythrin or lecanoric, gyrophoric, or alectoronic acids and yields pinks, purples, and fuchsias. Lichens historically used for these colors were *Ochrolechia* species, whose dye was called "cudbear," and *Roccella* species, whose dye was called "orchil." Because *Roccella* is very rare in North America, it should not be collected for any reason, least of all for dyeing.

Not all lichens work as a dye source. *Lichen Dyes: The New Source Book* by Karen Diadick Casselman names applicable species, which method to use, and gives detailed steps on the methods. Check online for workshops by Alissa Allen, who teaches the use of local fungi and encourages conservation of both lichens and mush-rooms. It is important to learn identification to avoid inadvertently collecting something rare just to see if it works as a dye source.

Although there are guidelines for ethical collection, the most sustainable approach is not to collect at all. Even the weedy species are best left alone. We live in a time when flocks of passenger pigeons no longer darken the sky. May our descendants never live in a time when tree trunks, tundra, and rocks are devoid of lichens. All the colors you can achieve from lichen dyes can be derived from other natural sources.

See also cudbear; lichen substances; litmus; orchil.

Etymology

The word "lichen" originated from Greek, was adapted into Latin, and then made its way into English (the same path taken by over half of all English words). Somewhere along the way, its definition split into a botanical meaning and a dermatological one, which continued to evolve in parallel.

The word started its journey as λειχήν, a noun derived from the verb λειω, "to lick." Its strict etymological definition would be "the licker." The earliest use of the word in surviving documents is by the dramatist and "Father of Tragedy," Aeschylus (ca. 525–456 BCE). He wrote of Apollo sending λειχήν as a punishment. In that instance, it meant a kind of blight that traveled the world destroying leaves. In another tragedy, he wrote of Apollo sending down λειχῆνας ἐξέσθοντας: the first word is the same lichen word in a different grammatical case, and the second is a qualifying participle best translated as "eating away." Together the two words are translated as "sores that eat away." Some translations are "what eats around itself." How we got there from "the licker" is probably obvious to a dog. The rest of us have to guess. Could it be that the lichen was "licking" its substrate and somehow consuming its surroundings? We may never know. The connection between leaf blight and sores on the skin is easily visualized. We also have to consider that dramatists are masters of metaphor.

Jump ahead a century to Hippocrates (ca. 460–370 BCE), who used the same word, λειχήν, the licker, in a medical context to mean ailments of the skin involving eruptions of papules (little bumps or warts). The anglicized version of the word, "lichen," has been carried forward into current-day dermatology with very specific meanings.

Jump ahead another century to Theophrastus (ca. 371–287 BCE). For the first time, the extant literature shows the word being used to describe what we know today as lichens (and probably including mosses and liverworts). It's not a giant leap to get from eruptions and bumps on the skin to crustose lichens with apothecia. It's as if the rock or the tree has a skin disease.

Roughly 300 years after the death of Theophrastus, lichens turn up in Roman records. Pliny the Elder (23–79 CE) and Dioscorides, his contemporary, both used the word to describe a lichen, and both suggested medicinal uses. They were not known to collaborate, but both drew heavily from the work of Theophrastus. In "Materia Medica," Dioscorides wrote: "Lichen grows on rocks and is also called bryon. It is a moss sticking to moist rocks. This is applied to stop discharges of blood, lessen inflammation, and heal lichen [papular skin disease], and applied with honey it helps jaundice. It also helps the fluids of the mouth and tongue [saliva]." He used the same word to mean both the lichen and the skin disease.

Despite the Aeschylus usage, the consensus among etymologists is that λειχήν did originally mean "lichen" or "moss" before it was carried over into the medical sphere as "rash" or "lichenlike skin growth." We don't have early enough botanical texts prior to Theophrastus to be certain. So many gaps—it's got me licked.

See also dermatology; Pliny the Elder; Theophrastus.

F AQs

Every field has its most frequently asked questions, and two that are always near the top of the list for lichens are "Do they harm my trees?" and "How do I get rid of them?" The short answers are "No" and "You don't need to do anything." Good news and good news.

Lichens need a place to attach where they have access to sunlight (not necessarily direct sunlight) and moisture, which they get mainly from the air. With no roots to penetrate the bark, they can't tap into the tree's vascular system and siphon nutrients. They manufacture their own food. They attach via filaments (rhizines), which do go into the bark, but not far. The removal of lichens from bark doesn't help the tree and increases the chances of damaging the bark, which does harm the tree.

Some lichen species need more light than others (old man's beard, for example), and so if you see an increase in light-loving species on older trees that have lost some

"What killed the tree?"
"Not lichens!"

branches and on dead trees, you may wonder if the lichens were responsible. The light came first, making a hospitable habitat, and the lichens followed. Lichen-covered trees, however, could also be hosting harmful fungi of the type that "eat" wood. Such fungi don't form lichens. Their lifestyle is independent of the lichens, although their effect on the tree could eventually let in more light by killing off some branches.

Lichens grow slowly, so you won't see them on young trees. This is another reason why people are led astray. "The trees were clean (!) when I planted them. It's only been five years, and now they have this stuff on them." That's more good news. Lichens don't do well in polluted air. If you have a lot of lichens on your trees, breathe deeply and give thanks to the environmentalists working on clean air regulations.

Fiction

The use of lichens as an agent in fiction is not a fertile field for novelists. John Wyndham, the author better known for *The Day of the Triffids* and *The Midwich Cuckoos*, is an exception. He combined his knowledge that lichens grow slowly, are long-lived, and are subject to overharvesting with his observations of the role of women in society, the short-term mentality of politicians, the worship of a youthful appearance, and the false claims of the cosmetics industry. What if something that slows the aging process could be extracted from a lichen? What if there were a skin care product that *actually* reduced signs of aging? What if the product had a known but undisclosed side effect that enabled people, starting with women, to live 200 to 300 years? What if supplies were limited and only the rich could afford it? What could possibly go wrong?

Wyndham's novel, published in 1960, is appropriately called *Trouble with Lichen*. Diana, his protagonist, has population growth and worldwide famine on her mind. "You know as well as I do that the world is in a mess and floundering deeper every day. . . . We are letting it drift toward that with an evil irresponsibility, because with our ordinary short lives we won't be here to see it. . . . There's only one thing I can see that will stop it happening. That is that some of us, at least, should be going to live long enough to be afraid of it for *ourselves*." Climate change, anyone?

Another novel, *The Collapse of Western Civilization* by the historians Naomi Oreskes and Erik M. Conway, published in 2014, is about climate change. The story includes a remedial role for lichens, which are known for their ability to grow in the harshest of conditions. It is the year 2393, the 300th anniversary of the "great collapse." The book's narrator, a scholar in the 2nd People's Republic of China, reports that back around the time of the collapse, a genetic engineer developed a fast-growing lichen with a superior ability to sequester carbon. "This pitch-black lichen, dubbed *Pannaria ishikawa*, was deliberately released from Ishikawa's laboratory, spreading rapidly throughout Japan and then across most of the globe. Within two decades, it had visibly altered the visual landscape and measurably altered atmospheric CO_2 . . . starting the world on the road to social, political, and economic recovery." Although *ishikawa* is an invented species of lichen, the authors chose a real genus, *Pannaria*, that contains nitrogen-fixing cyanobacteria, *Nostoc*. Lichens with an algal photobiont would also sequester carbon, but the needed algal partners would be hard to find on an overheated planet. Also, a damaged landscape would recover sooner with access to nitrogen; plus *Nostoc* is a possible food source. A cyanolichen is the better choice.

Wyndham thought of his work as "logical fantasy" rather than science fiction. I would call it social satire. To call the Oreskes-Conway book sci-fi minimizes the science. It is heavily science-based fiction. I would call it consciousness-raising and a "fable for tomorrow," in the genre of *Silent Spring* by Rachel Carson.

Fluorescence

Lichens glow. Not all, but some. In a process called "biofluorescence," chemicals convert high-energy ultraviolet (UV) light into lower-energy visible light. Of all the chemicals that lichens make (over 1,000), only some are photoactive (for example, lichexanthone and thiophaninic acid). It's known that UV radiation can harm photosynthesizing algae and that chemicals in the lichen cortex (the outer skin) act as a kind of sunscreen to protect the very important algae—the cooks in the lichen kitchen.

Many *non*-lichenized fungi glow in the dark owing to a different process called "bioluminescence." The difference is that luminescence is created by an enzymatic chemical process within the organism that releases energy as visible light. You can see it unaided in the dark. A mushroom example is the jack-o'-lantern. A well-known insect example is the firefly.

A long-wave UV black light (with a wavelength of 365 nanometers) is a useful part of a lichenologist's tool kit because the ability of a lichen to fluoresce and the color of the fluorescence are clues in the identification process—to say nothing of the joy that bioluminescent fungi can bring to a nighttime walk. Inexpensive battery-powered lights of the type used by rock collectors and other hobbyists serve the need and are sufficient for amateurs. Lichen colors that you see under UV light are quite different from the same lichen viewed in daylight. For example:

1. Typical green, gray-green, or whitish lichens may show up as vivid pink, blue, red, or orange under UV light. Some of these lichens are practically un-

detectable in daylight; a crustose gray matte lichen on the gray matte surface of Florida palmetto bark, for instance, is clearly defined in sharp-edged yellow patches under UV light. Similarly, *Pertusaria pustulata*, one of the wart lichens, is often close to the same color as the bark it is growing on; however, it is easily spotted at night because of its bright orange color under UV.

2. *Psilolechia lucida* is a Granny-Smith-apple-green crustose lichen that grows on high-humidity surfaces on the undersides of rocks in northeastern North America and west to the Great Lakes. It is easy to identify in daylight by its habitat, its color, and its powdery surface, but it also glows orange under UV.

In addition to fluorescence from chemicals in the fungal component of the lichen, chlorophyll in the algal component is capable of its own fluorescence, which provides a measure of photosynthetic productivity. Portable fluorometers are now available to measure and compare the productivity of different species (plants and lichens) at different locations in a habitat and at different heights in the canopy, all *in situ*.

Foliose

One of the three major growth forms of lichens is called foliose. The words "foliose" and the more familiar "foliage" are both derived from the same Latin root, *folium*, meaning "leaf," but "foliage" is used for plants and "foliose" for lichens.

Foliose lichens have lobes that are thin like leaves and have a distinct upper and lower surface. The lobes usually lie flat on the substrate. Anyone familiar with plant

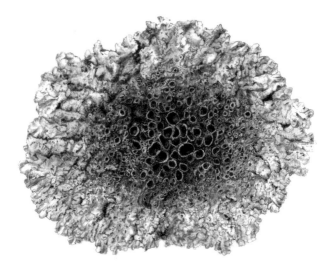

identification will know the importance of leaf shape, size, edge, color, ornamentation, attachment, etc. The same holds for foliose lichens. These physical character- istics help distinguish one species from another.

At the microscopic level, the tissue within the lobes is organized into distinct layers, most of which are fungal. The upper surface of the lobe is made of tightly packed fungal threads that form a cortex, or skin, which covers a layer of photosynthesizing cells (algae or cyanobacteria), below which is a layer of loosely packed fungal threads (medulla), and then usually at the bottom is a lower cortex. Fungal structures on the lower cortex, if present, hold the lichen to the substrate. When there is no lower cortex, threads from the medulla attach directly to the substrate.

Like the rest of the natural world, foliose lichens don't satisfy the human desire to fit them into clearly delineated categories. For example, a common type of foliose lichen with lobes so very small that they're not even called lobes but "squamules" is categorized under the subgroup descriptor of "squamulose." As some squamulose lichens (in the *Cladonia* genus) mature they lose their primary squamules and are left with vertical stalks that bear fruiting bodies. In this stage, they emulate the shrubby fruticose types. Other foliose species that don't stay flat on the substrate (for example, *Cetraria arenaria*, the sand-loving Iceland lichen) could also be confused with fruticose types, whereas some are so closely bonded that you can't be sure if they are foliose or crustose (such as the *Candelina* species, aka yolk lichens). Jelly lichens, such as those of the *Collema* and *Leptogium* genera, conform more closely to the typical foliose look on the exterior but don't conform to the stratification norm on the interior.

See also crustose; fruticose; thallus.

Food (for Humans)

Very few lichens are poisonous, but they are difficult to digest and bitter on the palate, so they are not part of the normal human diet. They offer little in the way of protein and fat. Most have vitamin C, but the primary nutritional contribution of the few species that are eaten is a relatively high percentage of starch (lichenin and isolichenin). A human tendency with starch-rich products is to grind them into flour and make bread, a practice that the ancient Egyptians up through current-day Scandinavians have used with lichens. Another tendency is to convert the starch to sugar and ferment it to produce alcohol. Many indigenous peoples have traditional methods for converting some lichen species into palatable, digestible, and even desirable food. Well-known lichens for human consumption are black tree-hair (*Bryoria fremontii*), Iceland moss (not a moss but the lichen *Cetraria islandica*), and rock tripes (*Umbilicaria* and *Lasallia*).

Lichens also provide food indirectly through the food chain. Just as spotted owls eat northern flying squirrels, which eat lichen, so humans eat caribou, which eat lichen. (Caribou and reindeer are the same species, *Rangifer tarandus*.) Caribou have lichen-specific enzymes (for example, lichenase) and microorganisms in their rumens that create absorbable sugars that are converted to meat, which, in turn, is consumed as a mainstay. Caribou also preprocess lichens for human consumption by the people who hunt them and have a strong no-waste ethic. When caribou (and sometimes musk ox) are killed in a winter hunt, their stomachs contain partially digested lichen. Some of it is eaten immediately while warm, some is saved and used as a dip, and some is stirred up with raw fish eggs into a concoction known as "stomach ice cream." Yum.

Old-time practices are dying out. Today's subculture of suburban foragers and weed-eaters is supported to a degree by specialized cookbooks with recipes for lichen-flavored custards, breads, and side dishes like rock-tripe with watercress. At the high end, restaurants, as exemplified by Noma in Copenhagen (rated number one of the world's top fifty restaurants five times since 2010, most recently in 2021), are incorporating lichens into haute cuisine as flavorings and adjuncts. No other culinary use is sustainable because lichens defy cultivation and grow too slowly to support anything but minimal collecting.

See also rock tripe; survival food.

Foolproof Four

Clyde Christensen, Professor of Plant Pathology at the University of Minnesota, coined the term "foolproof four" to describe four edible mushrooms that can be distinguished from others easily and safely, enabling beginners to gain confidence with their identification skills and begin to enjoy eating wild mushrooms. The four are morels, puffballs, sulphur shelf (aka chicken of the woods), and shaggy manes. The same approach can be applied to lichens, not for culinary purposes (as most lichens are inedible) but to satisfy a desire to call things by their name, as if greeting a friend.

Christensen's criteria for selecting the foolproof four mushrooms were that the species be relatively common, be unlikely to be confused with other species, have a common name, and be recognizable without a hand lens, chemicals, or microscope. I offer the following four lichens as examples of common species in northeast North America that meet similar criteria. Other regions and specialized habitats within a region will have a different set.

Pink earth (*Dibaeis baeomyces*) is a coarse, pale-gray crust that coats large areas of otherwise bare soil and is unmistakable when fertile. It is covered with little pink-headed apothecia on short white stalks. They look like tiny mushrooms. Pink earth can be so prolific that, from

10 mm

Pink earth

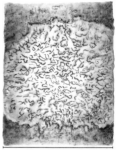

40 mm

Common script lichen

40 mm

Bushy beard lichen

4 inches

Smooth rock tripe

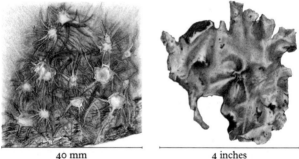

a distance, the patch of ground looks pink. It is common in areas of disturbed soil such as along roadsides, in old gravel pits, and under power lines. This lichen is one of the early colonizers of exposed soil, helping to hold it, improve it, and make it suitable for other things to grow. Hello, Dibaeis (rhymes with Tobias).

Common script lichen (*Graphis scripta*) is a pale-greenish-gray crustose lichen that grows on bark and looks like the runic markings in black ink of some forest-dwelling scribe. The black lines, which are modified apothecia, are fairly thin, irregular, pointed at the ends, and often branched (except on birch, where they follow the horizontal grain in the bark).

Smooth rock tripe (*Umbilicaria mammulata*) is a large foliose lichen that sometimes covers entire large boulders and rock faces in the woods. When dry, it is grayish brown, smooth, and very brittle. When wet, it is green and floppy and has a look of seaweed. The underside is completely black. Individual specimens range in size from two to twelve inches and occasionally are larger. They are attached to the rock at a single central point (umbilicus).

Bushy beard lichen (*Usnea strigosa*) is one of the many lichens in the genus *Usnea* that are all loosely referred to as old man's beard. It is a yellowish-green fruticose lichen (meaning that the thallus is shrubby), and the numerous side branches (fibrils) on the branches emphasize its bushy appearance. Just as a flowering plant may be hard to recognize when it isn't flowering, so too with the bushy beard when it's not fruiting. When it is fruiting, the very distinctive apothecia at the ends of the branches are quite large, cup-shaped, and pale orange in color, and they have rims adorned with green fibrils.

A good activity for local nature groups is to come up with their own foolproof four. As people learn how to look, the four will become eight, then twelve, and soon you have a local lichen club.

See also apothecium; old man's beard.

Franklin's Expedition

On May 23, 1819, Sir John Franklin, an officer in the British Royal Navy, set out on a journey to map parts of the northern coast of North America. Of a party of twenty, nine survived. To avoid starvation, they ate lichens and shoe leather to supplement the very occasional deer or partridge they shot. Sir John Richardson, a naturalist in the party, noted that the lichen "served the purpose of allaying the appetite but was very inefficient at recruiting strength."

The following excerpts are Franklin's own words from his *Narrative of a Journey to the Shores of the Polar Sea in the Years 1819, 20, 21, and 22* (published in 1823):

Sept 7, 1821: In the afternoon we got into a more hilly country, where the ground was strewed with large stones. The surface of these was covered with lichens of the genus *gyrophora* which the Canadians term tripe de roche. A considerable quantity was gathered, and with half a partridge each, (which were shot in the course of the day,) furnished us with a slender supper, which we cooked with a few willows, dug up from beneath the snow. We passed a comfortless night in our damp clothes, but took the precaution of sleeping upon our socks and shoes to prevent them from freezing.

Sept 13: We supped off a single partridge and some tripe de roche; this unpalatable weed was now quite nauseous to the whole party, and in several it produced bowel complaints.

Sept 16: We allayed the pangs of hunger, by eating pieces of singed hide. A little tripe de roche was also obtained. These would have satisfied us in ordinary times, but we were now almost exhausted by slender fare and travel, and our appetites had become ravenous.

Sept 18: No tripe de roche was seen to-day, but in clearing the snow to pitch the tents we found a quantity of Iceland moss [the common name of the lichen *Cetraria islandica*], which was boiled for supper. This weed, not having been soaked, proved so bitter, that few of the party could eat more than a few spoonfuls of it.

Sept 22: Our supper consisted of tripe de roche and half a partridge each.

Sept 23: They had picked up some pieces of skin, and a few bones of deer that had been devoured by the wolves last spring. They had rendered the bones friable by burning, and eaten them, as well as the skin; and several of them had added their old shoes to the repast.

Oct 10: There was no tripe de roche, and we drank tea and ate some of our shoes for supper.

Oct 12: I was unable to walk more than a few yards. . . . My other companions . . . went to collect bones, and some tripe de roche which supplied us with two meals. The bones were quite acrid, and the soup extracted from them excoriated the mouth if taken alone, but it was somewhat milder when boiled with tripe de roche, and we even thought the mixture palatable, with the addition of salt, of which a cask had been fortunately left here.

Franklin's last polar expedition was in 1845–1846. The two ships were trapped in the ice and all 129 men were lost.

See also rock tripe; survival food.

Fruticose

One of the three major growth forms of lichens is called fruticose. The other two, crustose and foliose, may be almost self-explanatory, but fruticose is a bit obscure. It is pronounced "froo-tuh-coze" and has nothing to do with fruit. The word comes by way of the Latin root *frutex*, meaning "bush" or "shrub," and the suffix *-osus*, meaning "abundance of." I took two years of high school Latin and remember texts about wars and empire building but not much about the natural world, with biological terms that would have been useful later in life. C'est la vie.

Fruticose lichens grow either upright and shrubby or pendant and droopy. Some are heavily branched and abundantly bushy; others are filamentous. The branches or filaments have no true upper or lower surface even though some are straplike. A cross-section through a

Ramalina americana

branch is circular or ovoid. The cells are arranged in rings, not top to bottom like foliose lichens. The outer ring, like bark on a twig, is a fungal cortex covering a layer of algal cells inside of which is more fungal material. The central core may be hollow or solid. Details of each layer contribute to the identification, as does the relative thickness of the layers.

Fruticose lichens are typically free of the substrate except for a single point of attachment from which the branching begins. Specimens can be either just about an inch tall or, in the case of the pendant Methuselah's beard, *Usnea longissima*, up to three feet, with an example in coastal Oregon of thirty-two feet.

People who build architectural or landscape models, and especially model-train enthusiasts, are probably familiar with a subset of fruticose lichens because they are widely used to represent trees and shrubbery realistically in miniature landscapes.

See also crustose; foliose; thallus.

Furbacken, Oscar (1980–)

Lichens have captured the attention of many artists. One of those is Oscar Furbacken, a Swedish artist born in Gothenburg in 1980 and now living in Stockholm. I have chosen to call him an ambassador for lichens because he renders these typically overlooked tiny wonders of nature in super-large format, making it possible for people to appreciate them, many probably for the first time. A half-inch-tall pixie cup *Cladonia* lichen on the soil at the edge of a path can live in obscurity, but not so a ten- or twenty-foot-tall sculpture of one silhouetted on the horizon. Similarly, an observer can't help but be drawn into the world of a sixteen-by-twenty-foot version of a sunburst lichen depicted in full fidelity on a gallery wall. Call this "lichens without a hand lens." If lichen appreciation is too much to ask, then at least Furbacken creates awareness. Thank you, Mr. Ambassador.

Furbacken also creates works with life-size lichens illuminated in cracks in massive concrete frames that focus attention on the content of the crevice. He is a close observer of microhabitats. The faithful attention to detail, whether super-size or life-size, would have been praised by the influential Victorian art critic John Ruskin, who shared Furbacken's fascination with lichens.

See also Ruskin.

Galápagos Tortoise

The iconic giant land tortoise of the Galápagos Islands stumbles into a lichen book because *Dirinaria picta*, the powdery medallion lichen, has been found growing on its carapace.

Dirinaria picta is found worldwide, in mainly tropical locales in both hemispheres; it favors the bark of palm trees growing along shorelines. In the Galápagos, it is very common on other substrates like rock, wood, and leaves. The high rounded dome of the tortoise's carapace, with its matte surface, is not too different from a boulder—albeit a boulder that moves (slowly). The land tortoise seems like an accidental lichen substrate that happened to work out, but only on male tortoises and only on the upper rear portion of their carapace. The lower areas on both sexes are often submerged (this is not an aquatic lichen), and the sides are continually scraped by vegetation as tortoises push their way through it, leav-

Size in relation to
five-foot woman

ing only the upper rear for the lichen to get a grip. Mating activity abrades any lichen trying to grow on a female.

The lichen and the tortoise share some lifestyle characteristics that allow them not only to cohabitate but have a good long relationship. Both are long-lived and able to survive long dry spells (up to a year for the tortoise). Both are connoisseurs of symbiosis. The lichen goes all out, living in its fully committed, no-way-back symbiotic state, while the tortoise has an arm's-length symbiotic relationship with two species of Darwin's famous finches. When finches approach the tortoise, it recognizes the advance as an invitation and extends its neck, stands tall, and lets the birds in to hunt for parasites in skin crevices it can't reach on its own.

See also substrates.

Glucose

Sugar beets, a common source of glucose, were scarce in Russia during World War II. Starch from grain and potatoes could be converted to glucose, but those sources were prioritized for alcohol production for military use, so a commercial process was set up to get glucose from lichens. The lichens were steeped in potassium carbonate to neutralize the bitter lichen acids and then heated with hydrochloric or sulfuric acid to convert the starch to glucose; the acid was then neutralized and the glucose solution filtered off and clarified before being either concentrated into a syrup (glucose molasses) or crystallized into a sugar. The lichens with the highest yields were witch's hair (*Alectoria*), Iceland moss (*Cetraria*), reindeer lichens (*Cladonia*), and old man's beard (*Usnea*).

Lichen glucose was deemed superior to beet sugar. Glucose molasses from Iceland moss was light in color and transparent, and it tasted of caramel.

The cost to harvest the lichen was low relative to beets, but the rest of the process was more costly. The operation ran for two years (1942–1943), proved the concept, and satisfied a short-term need, but harvesting lichen by the ton to feed commercial production was not sustainable.

See also alcohol.

Hallucinogen

In the northwest headwaters of the Amazon, various plant-based hallucinogens are important in the culture of most tribes, who use them medicinally, spiritually, and also socially. In 1981, two scientists, E. Wade Davis and James A. Yost, were doing ethnobotanical studies in eastern Ecuador when they came upon a small group, the Huaorani (or Waorani) people, who had other ideas. They thought hallucinogens were extremely antisocial, to be used only by shamans, and then only in secluded spots out of sight of everyone. They had two sources of hallucinogens: one was the woody vine closely related to the plant known as ayahuasca, or soul vine, and the other was a conspicuous bright green-blue lichen. The last remembered use of the lichen, by now over 100 years ago, was when a "bad shaman ate it to send a curse to cause other Waorani to die." Prepared as an infusion, the lichen (or perhaps remorse) gave the drinker severe headaches and caused confusion.

Davis and Yost collected a sample of the lichen, which turns out to be the only known specimen in the world. It is housed in the Farlow Herbarium at Harvard University. It was identified by the eminent twentieth-century lichenologist Mason E. Hale as a member of the genus *Dictyonema*. Filed under that name and preserved for posterity, it was available if and when needed for scientific analysis. As one of the most colorful lichens in the herbarium, it could be called psychedelic.

Jump ahead to the twenty-first century, when lichenologists working on tropical lichens, including the *Dictyonema* genus, took a closer look at the herbarium specimen. They determined that it was a new species and named it *Dictyonema huaoroni*, in honor of the Amerindians who reported its shamanistic use. Chemical analyses suggested the presence of tryptamine and psilocybin. Fresh specimens are needed to turn those suggestions into confirmation, but that would be difficult. Not only is *Dictyonema huaoroni* rare (Yost had known of it for seven years before he found it), but it is probably a canopy species. Its habitat is dense Amazonian forest threatened by oil exploration. The area is one of the world's hot spots for extremely high biodiversity. A lichen foray would be an intense trip, undoubtedly involving heightened awareness and severe headaches.

Over 200 species of mushrooms are known to contain psilocybin. Some research has suggested the presence of hallucinogens in lichens, but so far nothing conclusive has been found. But now the new studies are close to providing answers. In addition, circumstantial evidence collected via oral tradition, especially as indigenous ways of life dwindle, is important. In this case, the lichen's distinctiveness is likely to have made it memorable for both its appearance and its associated curse-making potency. Further, it is a basidiolichen in which the fungal partner is of the same type as the mushrooms that produce psilocybin.

See also basidiolichens.

Hand Lens

A hand lens is necessary to appreciate the many tiny and intricate structures of lichens. The fine detail is worth seeing for aesthetic reasons alone, but some detail is also necessary for identification. Hand lenses come in different magnifications up to thirty times (30x). For lichens, more is not better. Ten times (10x) is the most useful. Anything over fifteen times (15x) has a limited field of view and a very low depth of field. Five times (5x) or seven times (7x) magnification is a lot better than nothing, but you will miss some features.

If you are interested enough to look at tiny things, you will want a good-quality lens made of glass rather than

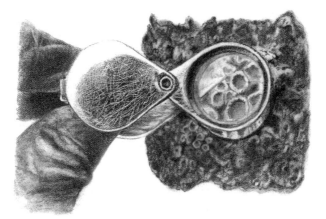

plastic, and one with built-in correction for spherical and chromatic aberrations. You might also consider a lens with an LED option. The downside is that it requires batteries— usually three tiny ones—though they last a very long time because ambient light is often enough. The upside is that you can get a good look at lichens in places like dark hemlock forests, bark crevices, and hollow logs.

Another option is the magnifier app available for both Apple and Android phones. So far, these apps are not ideal. The magnification is adjustable, and the phone flashlight provides extra light, but at ten times magnification, it is a challenge to hold steady. I expect the apps will improve with new phone releases, but for now, the hand lens is superior.

Once you have a decent lens, you might want to put it on a lanyard and wear it on your outings. Alternatively, you could make another naturalist happy when he or she finds the one you left in the field.

Herbivory

Lichens are old. They've been around for at least 250 million years—plenty of time to develop the herbivory deterrents necessary for their slow-growing, stationary lifestyle. The chitinous cell walls of the fungus and the many bitter and acidic lichen substances deter a lot of animals, but not all.

Moose, caribou, white-tailed deer, mule deer, elk, bighorn sheep, pronghorn antelope, and musk ox are among the large mammals (not counting humans) that eat lichens. Of those, caribou and reindeer have the highest percentage of lichens in their diet (around 90 percent in winter and 50 percent in summer). That is enough for a whole group of lichens to become known as reindeer lichens, most of which grow on the ground. When they are covered in snow, as is often the case, tundra caribou know how to find them and kick holes in the snow to get to them. Woodland caribou, moose, deer, and elk eat other species of lichens that grow on bark, especially the filamentous ones known as witch's hair, horsehair, and tree-hair. In parts of the boreal forest, there is a lichen graze line showing the reach of these animals.

Smaller mammals also eat lichens. Black snub-nosed monkeys (aka Yunnan snub-nosed monkey), who live at elevations up to 13,000 feet in the coniferous forests of southern China (where there is snow for much of the year), pluck lichens off the branches, hold them with both hands, and tear bits off like a child eating cotton candy. Rabbits and voles nibble on bark-dwelling lichens in winter. Northern flying squirrels depend on lichens for the bulk of their food, and some have developed an interesting larder plan. They build their nests with the

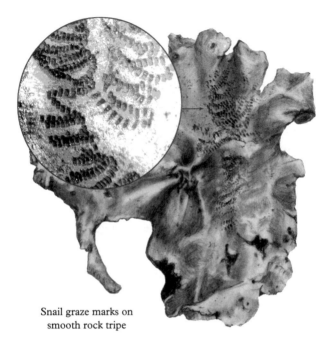

Snail graze marks on
smooth rock tripe

same horsehair lichens (*Bryoria* species) that they eat.
At the end of the winter, when the nest is no longer
needed, they eat it.

Animals that eat bark eat lichen by accident. In eastern Massachusetts, I have watched beavers in late fall
and early winter eating every morsel of bark from black
birch branches that were hosting crustose lichens and
small colonies of foliose lichens.

Many invertebrates—mites, caterpillars, springtails,
slugs, snails, earwigs, tree cattle, and black termites—
consume lichens. Some are large, like the Christmas

Island red crabs in the Indian Ocean, but most are tiny and leave no trace. The caterpillars are often highly camouflaged, but slugs and snails leave distinctive tracks that are easy to find. They don't eat far down into the body of the lichen; they scrape away the fungal cortex and graze on the layer of green algae just below. Then, instead of going deeper, they move along and scrape again. This leaves a pattern of scrape marks where each scrape is exactly the width of the snail's radula. It is most obvious on lichens with smooth flat surfaces, like rock tripe. Once you have seen the pattern, you will know it always.

In addition to providing food for wildlife, lichens are harvested by farmers in Iceland and northern Scandinavia as supplemental fodder for pigs, cows, sheep, and goats.

Hummingbird Nest

The ruby-throated hummingbird is the only hummingbird that nests in North America east of the Mississippi River. It is remarkable in many ways, and so too is the precisely engineered nest it builds. The nests are so small and so well camouflaged with lichen that it takes an element of luck to find one. If you spot a male ruby-throat doing the giant U-shaped dives of his courtship display, and if you see him switch to rapid side-to-side flights, you may eventually find a nest in the area. Watch for the female, who builds the nest. She usually chooses a place at least ten feet above the ground in a deciduous tree. Over the next six to ten days, she will construct a tall, cup-shaped nest of soft fibers like milkweed or cattail fluff, using spider silk, which is suitably strong, sticky, and stretchy, to hold it together. She reinforces

the exterior of the nest with bits of lichen—like tiles in a shower stall—that are also held in place with spider silk.

The nests I have seen were completely coated with fragments of the very common greenshield lichen *Flavoparmelia caperata*. The lichen pieces are meticulously attached wall-to-wall over the whole exterior and the rim so that the soft interior is completely protected. Most amazing is that every single lichen piece is positioned with the upper (green) cortex of the lichen facing outwards and the lower (black) cortex facing in, so that the lichen provides camouflage in addition to structure. It looks like a natural clump of lichen on the branch. Another bit of ingenuity is that the elastic nature of spider silk allows the nest to expand (within limits) as the hatchlings grow.

If you happen to notice a ball of lichen on the ground, small enough to sit in the palm of your hand, take a closer look. It might be a dislodged hummingbird nest.

See also camouflage.

I celand Moss

Cetraria islandica is incorrectly called Iceland moss. It's a lichen. Its Icelandic name is "fjallagros," which means mountain grass, another misnomer. It grows on the ground, mainly in arctic and alpine tundra zones, not only in Iceland but also in many other countries. It is one of the most common lichens used for human food along with black tree-hair (*Bryoria fremontii*) and rock tripes (*Umbilicaria* and *Lasallia*). Most modern lichen books have dropped the "moss" name and are calling it *true* Iceland lichen to distinguish it from other *Cetraria* species. A close relative, *Cetraria ericetorum*, is known simply as Iceland lichen.

Cetraria islandica, which contains a high percentage of the lichen starches lichenin and isolichenin, has a long history of use as food in Iceland, Norway, and Sweden dating back to the Vikings. The lichen was washed, boiled, dried, and ground into a flour, mixed with wheat flour, and baked into bread and ship's biscuits. A common name related to this use in Norway was Brödmose, or bread moss. The lichen flour was sometimes boiled with milk until it formed a jelly, which was the basis for a variety of custardlike desserts. Some lemon juice, sugar, chocolate, and almonds were added to the dried jelly to make candy. The lichen was also cooked into a gruel, or porridge, which would be mixed with oil, egg yolk, and sugar to make a dish that even "the most pretentious person will like."

This lichen continues to be used as food in a limited way, mainly for making flatbreads and a gruel called Iceland moss soup (Fjallagrasamjolk). This soup is a simpler version of the old gruel: its only ingredients are lichen, milk, and sugar, with fruit or other additives as desired.

The lichen's traditional use as medicine for easing sore throats, coughs, and ulcers and strengthening the immune system also continues today, largely for throat and respiratory ailments. You don't even have to collect your own: *Cetraria islandica* is available commercially—clean, dry, and ready to make into a hot-water infusion.

Isidium (*plural* Isidia)

Vegetative reproduction by special propagules is of huge importance in lichens. It allows the lichen components to stay together on dispersal. Isidia, one of the more common types of propagules, are tiny protrusions on the lichen surface designed to break off easily and start a new lichen colony. They contain both fungi and algae

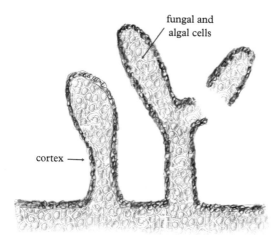

fungal and
algal cells

cortex →

and perform the same vegetative reproductive function as soredia (powdery propagules) but are usually sturdier and smoother because the lichen cortex remains intact. They may be constricted at the base to facilitate their breaking off. Like soredia, their shape, location on the thallus, and presence or absence are clues to identification. They are visible to the naked eye—what you may perceive at first is a change in color on the thallus, especially toward the center. A hand lens is really helpful to discern their shape and confirm that they are not soredia.

Some isidia are simple fingerlike projections, some are more like clubs, and others are globular, flattened, branched, or coralloid (resembling coral). To complicate matters, sometimes the cortex breaks down and you are left with "naked" isidia. In the wonderful world of lichens, the opposite can happen. Undispersed, stay-at-home soredia may develop a cortex and become isidia look-alikes. Flexibility.

See also soredium.

Lace Lichen

The first lichen to be recognized as an official state symbol in the United States was lace lichen, also known as fishnet lichen. State symbols originated at the 1893 World's Fair in Chicago when a National Garland of Flowers with a representative flower for each state was exhibited. The exhibit started a trend. Soon there were state birds, state trees, state insects—even state cookies. The collection of symbols for any state is meant to represent its culture and nature. In 2015, California became the first state to recognize a lichen as an official state symbol.

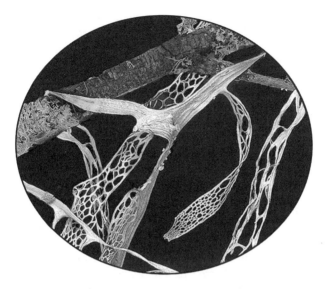

Section I, part of the bill to create a state lichen, reads: "*Ramalina menziesii*, commonly known as lace lichen, is a common lichen found throughout much of California from the northern to the southern border of the state, and as far as 130 miles inland from the coast. Naming *Ramalina menziesii* as the official state lichen of California will help promote appreciation, education, and study of lichens in this state." It was signed by Governor Jerry Brown on July 15, 2015.

They chose a very fine lichen. Edward Tuckerman wrote in 1872 that it is "one of the most remarkable of the species characteristic of the West Coast." Lace lichen has the advantage of being easily recognized by people not familiar with lichens. The gray-green pendent branches usually grow on oaks and can become a meter long. Other lichens have that growth form, but this is the only one to make you think of lace or fishnet. Close to the coast, the lichen branches tend to be very narrow, with the lacy portions only at the tip. Inland populations have wider branches and extensive areas of laciness on side branches. It grows quickly (for a lichen) and is plentiful enough in some areas to provide an important food source for sheep and deer—though not so plentiful that mule deer won't occasionally fight over it.

Lace lichen held magical properties for the Kawaiisu people, who used it in attempts to manage the weather. Putting it in water could make it rain, and burning it could prevent thunder and lightning. The magic had at least a couple of problems. It wasn't dependable—its power was all in the timing. (Same problem I have with rain dances.) Another problem was that it was widely available, which impinged on job security for the rain shamans.

The more than 200 species of the genus *Ramalina* are distributed worldwide, though some have a limited range. Lace lichen, for example, is found only in a strip of land running from Baja California up through southeastern Alaska.

See also Tuckerman.

Lichen Moths

A tribe of moths whose caterpillars feed on lichen are called lichen moths. Some of the adult moths, like the black-and-yellow lichen moth, painted lichen moth, and scarlet-winged lichen moth, are boldly colored. Tiger moths, members of the same subfamily (*Arctiinae*), are just as boldly marked and rely on aposematism for survival. Other adult lichen moths, like the little white lichen moth and the bicolored moth, rely on camouflage as a survival strategy.

The lichen-feeding caterpillars of these moths have a different way to discourage predation. They are spiny or hairy, and some of the spines have barbs. Think of a mouthful of a woolly bear caterpillar—or the time your dog had a tussle with a porcupine. The first prize for creative pilosity goes to the Australian lichen moth, *Cyana meyricki*, which is highly deserving of, but still waiting for, a common name. When the caterpillar is ready to pupate, it pulls out its extremely long hairs and constructs a cage within which it suspends itself.

Bitter compounds in lichens discourage many critters (including humans) from eating them, but caterpillars

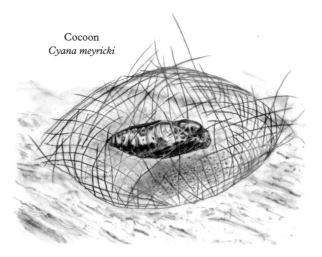

Cocoon
Cyana meyricki

of the brightly colored adult lichen moths are not deterred. They have evolved not only to feed on lichens, thus making themselves distasteful to predators, but also to retain some of the bitter phenolic compounds through pupation and pass them on to the adult moth. The disagreeable taste of all the life-cycle stages (caterpillar, pupa, and adult) discourages bats, birds, and even parasitoid flies and wasps. The strategy seems to be successful given that lichen moths are common.

Little has been published about the caterpillars' lichen preferences. The black-and-yellow lichen moth has been successfully raised in a lab on a diet of mealy rosette lichen (*Physcia millegrana*), a common foliose lichen with very narrow lobes, and the scarlet-winged lichen moth has a fondness for the lichens on red pine.

See also aposematism.

Lichenicolous Fungi

Lichens grow on many different substrates, and they themselves are occasionally a substrate. Lichenicolous fungi are fungi that grow on lichens. The suffix "-icolous" means "living or growing on." It turns up a lot in lichen terminology. Lignicolous lichens grow on wood (*lignum* is Latin for "wood"), saxicolous lichens grow on rocks (*sax* is Latin for "rock"), corticolous lichens grow on bark (*cortex* is Latin for bark), and so forth. The fungi that *become* lichens are not lichenicolous—they are *lichenized*.

Lichenicolous fungi are typically very small. A casual observer, who might not notice them at all or mistake them for part of the lichen, shouldn't feel bad. Until recently they were mostly overlooked by lichenologists too, but now over 2,000 species have been named, with at least another 1,000 expected.

Most lichenicolous fungi are host-specific parasites feeding on living lichens; some are saprophytes feeding on dead lichens. The parasitic ones may or may not kill their host. The strangest lichenicolous fungi are a small group of lichen wannabes. They live for a while on a lichen and then steal its algae and become lichens themselves. The cowpie lichen, *Diploschistes muscorum*, starts life as a non-lichenized fungus living as a parasite on species of *Cladonia*. It incorporates algae from its host, gives up parasitism for mutualism, and lives on as a lichen. A reformed parasite?

Lichenometry

Lichens grow slowly and can live a long time, some possibly up to 4,000 years. Lichenometry is the technique of using a lichen's rate of growth to calculate the age of an existing thallus and thereby deduce that the surface it's growing on must have been available for at least that many years. The technique has been used on moraines to understand the rate of glacial retreat. (Today a stopwatch would do, but lichenometry enables a view into history.) It has also been used for dating earthquake faults and the Easter Island statues.

Easter Island moai

The tricky part of the technique is coming up with the lichen's growth rate. Not all lichens grow at the same rate, and the rate varies from year to year. The species being measured, almost always a crustose lichen, has to be calibrated against old, large thalli of the same lichen on substrates of known age. Despite the difficulties, the technique is useful when carbon dating and dendro-chronology are not applicable.

Lichen Substances

Lichens are chemical factories. As autotrophs, they have to produce the compounds necessary for life (proteins, amino acids, carbohydrates, vitamins, etc.). These "primary" or intracellular compounds are not unique to lichens. What makes lichens interesting (well, one of many things) are the hundreds of "secondary" extracellular metabolites. There are over 1,000, and all but about 60 occur nowhere else in nature. The fungal partner produces them and deposits them on the outer surface of its cells. Some are placed only in the upper skin (cortex) of the thallus, where they have one or more functions: protecting the algae below from damaging amounts of light, discouraging browsing with their nasty taste, or acting as antifungal and antibacterial agents. Others are placed only in the interior (medulla) of the thallus, perhaps because they repel water and help maintain the air spaces needed for efficient gas exchange in photosynthesis.

The few lichen substances that are not colorless give the thallus a characteristic color. For example, usnic acid, which is yellow, is present in yellow-green foliose lichens, which are often seen on tree trunks side by side with gray-green foliose lichens, which contain no usnic acid. Firedot lichens (*Caloplaca*) and sunburst lichens (*Xanthoria*) get their orange color from anthraquinone pigments, often parietin.

Lichen substances are part of a lichen's identity. Interested amateurs who want to take their identification skills beyond morphology can do chemical "spot" tests with household products (lye and bleach) to confirm the presence or absence of groups of compounds. The professionals use spot tests too, along with an array of advanced methods like thin-layer chromatography, high-performance liquid chromatography, and now DNA analysis.

Human uses of lichens for dyeing, fixing perfumes, and a variety of medical applications are all dependent on these substances. Research continues in the hopes of finding antiviral and anticancer drugs, but harvesting enough lichens for this purpose is hard: lichens are difficult to cultivate, and they grow very slowly in nature. When the fungal component of a lichen is grown separately from its algal partner, which is an artificial situation, it produces only tiny amounts of lichen substances—or none. Many mysteries remain.

See also Nylander.

Linnaeus, Carolus (1707–1778)

Carolus Linnaeus, aka Carl Linnaeus, was a Swedish botanist most noted as the father of modern taxonomy. In 1735, he published "The System of Nature" (*Systema Naturae*). This work, only eleven pages long, introduced a hierarchical ranking system designed to classify all of nature. Everything was to be assigned to one of three kingdoms: plants, animals, or minerals. The kingdoms were divided into classes, orders, genera, species, and varieties. A uniform naming system, a unique two-part Latin name made up of genus and species, was to be assigned to every living thing. The twelfth and final version of *Systema Naturae*, published in 1766–1768, had 2,400 pages. Many changes and additions have been made since then, but the basic principles and the binomial method of naming species remain.

Linnaeus named around 7,700 species of plants. He had become very interested in the sexual reproduction of plants. He called the stamens "husbands" and the pistils "wives." Their number and location informed his taxonomic decisions. The "husbands" determined the class and the "wives" determined the order.

Linnaeus had less time for lichens. He named 109 species and assigned them all to a single genus he called *Lichen*. The kind of man who called a spade a spade.

It was up to Erik Acharius and other followers of the Linnaean system to create the beginnings of a proper lichen taxonomy. All was well for 100 years. Then, in 1867, Simon Schwendener showed that lichens exist as a combination of fungi and algae, a situation outside the scope of the Linnaean requirement of individuality. Furthermore, lichens (and all fungi) were prisoners in the kingdom of plants. Fungi finally got their own kingdom in 1969—a rate of change comparable to lichen growth rate.

Right up into the twenty-first century, Linnaeus was scorned for dismissing lichens as "the poor trash" of vegetation. Big sorry, Carl, from all the lichen people. This was a mistake caused by carrying forward a botched translation of the Latin phrase *pauperrimi rustici*. If they had had free online translation apps in the nineteenth century, they would have known that it means "poor peasants." Some translator may have had a low opinion of peasants, but not Linnaeus. Although he had little time for lichens taxonomically (his poor peasants had no flowers), he appreciated the work they did in colonizing rocks and creating habitat for plants. He wrote that lichens "prepare the ground for all plants and are for that reason, in spite of their modest appearance, quite remarkable in the economy of nature."

Around 1900, a group of biologists came up with the idea of linking binomial names with "type specimens." The specimen would serve as the standard for the species and allow determination of whether a new, as-yet-unnamed species had been found. In 1959, the British botanist W. T. Stearn designated Linnaeus's remains as the type specimen for *Homo sapiens*.

See also Acharius; naming; Schwendener.

Lips

Yes, lichens have lips. If you're being formal, you would call them lirellae, from the Latin *lira* for "furrow" and the suffix *-elle* for "small." Michel Adanson (1727–1806), a French botanist and naturalist, coined the name to describe the elongated fruiting bodies on lichens in the genus *Graphis*.

The lips are the walls of the fruiting body and, like human lips, they taper at the ends. A "small furrow" along the center of the fruiting body creates the appearance of slightly parted lips. This is the mouth through which the spores are ejected. The opening may be less than half a millimeter wide, but that central furrow is

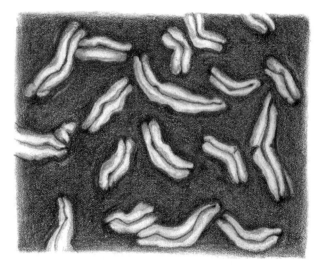

Powdery script lichen *Dyplolabia afzelii*

always there. If you take an elastic or rubber ring and stretch it out in opposite directions a little or a lot, you will have an idea of the type of shapes. The lips may be black or white, thick or thin, smooth or cracked. For example, the powdery script lichen, *Dyplolabia afzelii*, has thick white lips; a woodscript lichen, *Xylographa rubescens*, often has cracked lips; the burnt lips lichen, *Phaeographa inusta*, has almost closed black lips.

Lichens with lirellae occur in other genera besides *Graphis* (such as *Allographa*, *Opegrapha*, *Phaeographa*, and *Xylographa*). These lichens are collectively referred to as script lichens, or graphid lichens, from the Greek *graphikos* for "writing." Not all lirellae resemble lips. Some are branched, some are clustered, and some (especially some of the black ones) look like scribbles or runes. If you find a lichen that looks like it's trying to tell you something—either in writing or by inviting you to read its lips—chances are good that you are looking at a graphid. Lichen identification can be fun—it's a matter of knowing how to look and interpret the signs.

Litmus

Litmus is a water-soluble dye extracted from lichens. The word is derived from the old Norse words *litr*, meaning "dye," and *mosi*, meaning "moss"—yet another situation where lichens were called "mosses" (like the misnamed Iceland moss and reindeer moss).

We all learned in chemistry class that litmus is an indicator of pH: red on the acidic side of neutral and blue on the alkaline side. What we didn't learn is that it was Arnaldus de Villa Nova (ca. 1240–1321), a renowned Spanish physician to three popes and three kings, who first discovered that litmus turns red in acid. He was also a theologian (in a time and place where innovative ideas could cause trouble) and an alchemist (a practice in which innovation is a necessity).

Litmus is usually extracted from the species in the *Roccella* genus collectively known as orchil lichens. They are the source of a purple dye called orchil (aka archil), which was used in ancient times by Greeks, Romans, Egyptians, and Phoenicians. It is still used as a pigment for artists. *Roccella* species are rare in North America. They grow in the Mediterranean region, the Canary Islands, Europe, South America, and Madagascar. Over the years they have been collected by the ton to support dye and litmus production, mostly dye. This practice is not sustainable for either industry, and these days other methods are available that not only measure pH but do it more precisely.

Litmus is not a single compound but a mixture of ten to fifteen different lichen substances. For the organic chemists, the compounds include azolitmin, erythrolitmin, spaniolitmin, and erythrolein, all of which are related to derivatives of phenoxazine. The extraction process involves fermenting the lichen in a sodium or potassium carbonate solution in the presence of ammonia. It goes through some color changes before it turns blue, at which point paper can be impregnated. For red litmus paper, a small amount of acid is first added to the litmus extract.

Six centuries after Arnaldus's discovery, the phrase "litmus test" has taken on a figurative meaning: a litmus test is a single factor by which the whole of something or someone can be judged. Example: a litmus test for a book on lichen lore might be that it include an entry for "litmus."

See also dyes; lichen substances; orchil.

Lizard Semen

We have long puzzled over the origin and nature of lichens, and scientists continue to work on the puzzle even as they find and name new species. Indigenous peoples all around the world have also long been familiar with lichens; in discovering uses for them, they assigned their own descriptive names based on the use, appearance, habitat, or substrate.

The Paiute people of western Nevada discovered that some yellow and orange rock-dwelling lichens in their area had antifungal and antibacterial properties. These lichens were worth knowing and therefore worth naming. They called these lichen splotches lizard semen. The imaginative name is based on two observations. First, western fence lizards, also known as blue-bellies, common throughout Nevada and neighboring states, sun themselves on paths, fence posts, and other elevated sunny spots, including the rocks where the lichens grow. Second, the lizards have a behavior that looks like little push-ups on the rocks.

The lichens are most likely either sunburst species (genus *Xanthoria*) or firedot species (genus *Caloplaca*), both of which contain anthraquinone pigments, some of which have exhibited antimicrobial activity in laboratory experiments.

Lungwort

Lungwort, *Lobaria pulmonaria*, is a favorite of mine. An art-collector friend once walked me through his collection prefacing his presentation of every piece with "This is one of my favorites!" Lichens can do that to you too. They are works of art in every sense. Any of the color plates in Brodo, Sharnoff, and Sharnoff's *Lichens of North America* (2001) could hold its own in a modern art gallery.

First off, lungwort is a large lichen. The thallus can cover extensive areas on tree trunks, and when it's moist, it's a vivid green. The lobes are broad (*Lobaria* is from the Greek for "lobe"), with two-way or three-way branching; squared off at the tips, they tend to curl out from the substrate. The surface of the lobes is highly textured with ridges and pits. Lungwort grows on tree trunks in mature

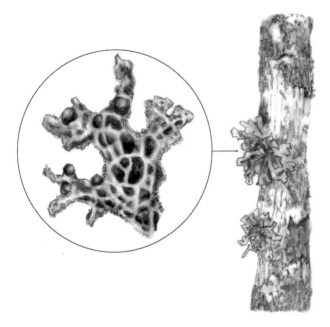

and unpolluted forests. At one time it was considered an old-growth-forest indicator.

Next, it has two methods of photosynthesizing. It is one of the 10 percent of lichens that use cyanobacteria (previously known as blue-green algae), but it doesn't put all its eggs in that blue-green basket. Its primary photobiont is the green alga *Symbiochloris*. The secondary photobiont, the cyanobacterium *Nostoc*, is super-important in old forests because of its ability to fix nitrogen. Lungwort is a relatively fast grower and is loosely attached to its substrate, so a lot of it ends up on the forest floor, where it contributes significantly as a natural fertilizer. Craft dyers pick up dislodged lichens in the mistaken belief that once fallen they have no ecological value and can be harvested at will. Not so with these nitrogen fixers. The forest still needs them.

Lungwort has three methods of reproducing. It has both of the common vegetative propagules (soredia and isidia), and it has apothecia where the fungal partner makes spores after sexual reproduction. Soredia and isidia are positioned on the lobe edges and the tops of the ridges emphasizing the sculptured surface of the thallus.

Lungwort is clearly an overachiever that deserves a new common name. The current name is based on its resemblance to human lung tissue (*pulmonaria*, from the Latin, means "of the lungs"). This resemblance led to its medicinal use for lung diseases, as suggested in the Doctrine of Signatures. Botanists already use "lungwort" as a name for flowering plants in the genus *Pulmonaria*, where the old English *wort* meaning "plant" is applicable. But lungs are not the first thing that comes to mind when you see the lungwort lichen. Some specimens resemble moose antlers. Erin Tripp and James Lendemer, in their *Field Guide to the Lichens of Great Smoky Mountains National Park*, suggest calling it "crown jewel of America"—they like this lichen even more than I do. "Lobaria pulmonaria for president!," they say. Gitksan people of the First Nations of British Columbia call it *nagaganaw*, meaning "frog dress." I could go with that.

Its rhyming binomial *Lobaria pulmonaria* is catchy. I see it in a musical to popularize lichens, costarring *Dimeleina oreina*. Where are you, Lin-Manuel Miranda?

See also Doctrine of Signatures; foliose; isidium; soredium.

Mackenzie, Elke (1911–1990)

Mackenzie, Elke (1911–1990)
This noted lichenologist made major contributions in the field of Antarctic lichens and in the devilishly difficult genus *Stereocaulon*. Her work is published under her birth name of Ivan Mackenzie Lamb. She took the name Elke Mackenzie in 1971 after gender reassignment surgery.

Mackenzie was born in England and educated in Scotland. She worked under the direction of Annie Lorrain Smith at the British Museum of Natural History, building an interest in and working on the collections of lichen flora of Antarctica.

During World War II, Mackenzie joined the secret Royal Navy Antarctic expedition dubbed Operation Tabarin, with a primary role as botanist and also serving as dog-driver and surveyor's assistant. The purpose of the mission, which allowed for scientific research, was to solidify British sovereignty on the continent. Fourteen men left from the Falkland Islands for Antarctica at the end of January 1944 and remained there through January 1946. Mackenzie was well regarded under the trying conditions for empathy and generosity and for encouraging an interest in lichens in others.

Of the 1,030 specimens collected on the expedition, 865 were lichens, of which one remains the only species known to survive when permanently submerged in saltwater. A few lichens live in semi-aquatic habitats, but only two are known to live full-time underwater. McKenzie's discovery, *Verrucaria serpuloides*, forms black patches at the base of submerged rocks as much as ten meters below mean high tide. (The other is the freshwater species *Peltigera hydrothyria*.)

After the war, Mackenzie took jobs in Argentina and Canada before becoming Director of the Farlow Herbarium at Harvard University in 1953, a position she held until her retirement in 1972. During that time, she did painstaking work on *Stereocaulon* (foam lichens, or as some call them, snow lichens), making major contributions to the body of knowledge.

Her legacy, beyond the enduring work on *Stereocaulon* and forty-three scientific papers, are two lichen genera, *Lambia* and *Lambiella*, five species in other genera, and Cape Lamb on Vega Island at the tip of the Antarctic Peninsula, named in her honor.

Only a person deeply acquainted with life's difficulties would have taken on the task of clarifying the *Stereocaulon* genus.

See also Smith.

Manna

Manna might be best known from the biblical legend of food that fell from the sky, as if from God, to nourish the Israelites in their exodus from Egypt. In the account from the Old Testament, Numbers 11:5–6, "We remember the fish which we ate freely in Egypt, the cucumbers, the melons, the leeks, the onions, and the garlic; but now our whole being is dried up; there is nothing at all except this manna before our eyes!"

Today "manna from heaven" refers to any unexpected gift in a time of need. The name is used in Sicily to describe the sweet exudate from ash trees. In the hot summer sun, the drips solidify into a sugary substance. Researchers have tried to explain the original manna in the desert wilderness, and the most likely candidate is a lichen. The evidence is, of necessity, circumstantial. Much of it is based on several occasions throughout the nineteenth century in the Middle East and western Europe when small whitish spheres of stuff appeared, some of it falling from the sky. A collection of this mysterious material was examined at the University of Brussels in 1893 and found to be the lichen *Lecanora esculenta*, now known as *Circinaria esculenta*.

Most lichens are firmly attached to their substrate, but a few, like *Circinaria esculenta*, have weak attachments or none at all and can become "vagrant." *Circinaria esculenta* grows in arid areas of Eurasia and North Africa on the ground, in sand, and in deserts in the shape of small, firm, rounded objects that are easily blown by the wind and rolled into piles, where they soften in the heavy morning dew. These piles can be swept up in whirlwinds and carried long distances, eventually showing up elsewhere as if by providence.

In one episode, amid a famine in 1829 on the southwest shore of the Caspian Sea, the appearance of this lichen was noticed when starving sheep were seen to be eating it avidly, giving people the idea to try it themselves. Manna from heaven! Anything was worth a try—they made it into bread. In another account, a heavy manna fall occurred in Turkey in 1890. This instance is thought to have been due to a tornado. Again, the lichen was made into bread.

Henri Chastrey's article "The Manna of the Hebrews," published in *La Nature*, a French weekly journal of popular science, in October 1898, reported that the lichen is known and well-appreciated by the nomadic Arabs. This substance has often saved them from famine, and hence death. At times the Arabs stockpile it. Collection is easy since the lichen does not attach to any foreign object, but appears to have been thrown onto the sand. . . . It has a starchy taste and a slight but pleasant sweetness. All herbivores, camels in particular, are very fond of it. The Arabs boil it in water and obtain a gelatinous paste, which the desert gourmets use in various ways. To conserve it they let it dry in the shade, where they wrap it, either with camel bladders or with sheepskin, once it has been reduced to a paste.

"Manna" remains an imprecise term when applied to lichens. It probably encompasses a number of desert-dwelling vagrant species. Common names like fat of the earth, honey earth, pigeon dung, and Ousseh-el-trab (excrement of the earth) indicate disagreement on its desirability as food. That *Circinaria esculenta* was in fact the manna of the Bible remains dubious. A benevolent God could have done way better.

Moss
True moss is a little, nonvascular, spore-producing plant whose name is recklessly thrown around and applied to plants that are not mosses and to lichens, which, of course, are not plants. The 12,000 species of true moss do *not* include the so-called Iceland moss, reindeer moss, beard moss, or oak moss, all of which are lichens; nor do they include the club mosses (spore-producing, nonflowering vascular plants), Spanish moss

(a flowering plant, not a moss, and not from Spain), or Irish moss (a red, branching seaweed also known as carrageen, found on many Atlantic shores besides Ireland).

Iceland moss has triple trouble in the naming department. It's a lichen, not a moss. "Iceland" is limiting since the lichen has an extremely large range that

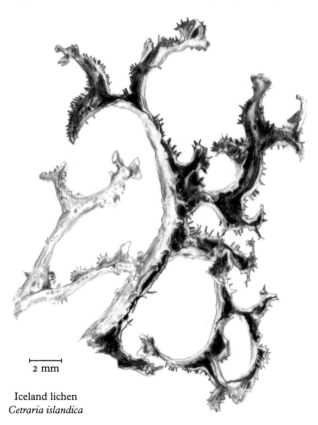

2 mm

Iceland lichen
Cetraria islandica

includes many countries in both hemispheres. Further entrenching the moss name, Scandinavians call it bread moss (*brødemose*), because in hard times it was ground up and added to wheat flour for bread-making.

Spanish moss, neither a moss nor a lichen, does have a lichen connection. Its scientific name is *Tillandsia usneoides*. *Usnea* is a lichen genus and the suffix "-oides" means "resembling." The plant is named appropriately for its resemblance to lichens in the genus *Usnea*.

The "idle moss" that Shakespeare mentions in his farcical *Comedy of Errors* is now thought to be lichen. If so, he can be forgiven. The play was first performed in 1594, when lichens were not only considered plants but were often confused with mosses.

The reason this mossiness persists is that old habits die hard. It was not until the mid to late nineteenth century that the true nature of lichens was made known by Simon Schwendener. Up to then, they were thought to be a puzzling form of lower plant life and "moss" had been a convenient bucket to drop them in. It's been over 150 years since Schwendener's discovery, and lichenologists have been diligently using the names Iceland *lichen*, reindeer *lichen*, etc., in books, but to no avail. We know they are not moss, but deep-seated, culture-based names are as firmly fixed as crustose lichens on rocks.

See also Schwendener.

Mummification

Imagine you are the director of collections at a lichen herbarium and you have seen thousands of packets with the typical annotations of date, location, identification, and collector name. One day when examining old collections typically from the nineteenth and twentieth centuries, you come upon a packet that says the lichen is from 500–800 BC and was found "within wrappings of a Mummy." This happened to Michaela Schmull, Director of Collections, Harvard University Herbaria and Libraries, who tracked down the story.

The sample had come to the Farlow Herbarium at Harvard from Edward Tuckerman's collection after his death. Tuckerman was a prominent nineteenth-century lichenologist to whom the lichen had been sent for identification from the Natural History Society of Montreal, where it had been found in 1859 during the unwrapping of a mummy. This was a time of great public interest in mummies, which gave museums prestige, and their unwrapping could be used to attract crowds, recruit new members, and raise money. X-rays had not been discovered, so unwrapping mummies had some scientific merit. The particular mummy was a female from Thebes whose name may have been Abothloé. She was in a highly decorated coffin and very well preserved. Sir J. W. Dawson, principal of McGill College in Montreal and acting president of the Natural History Society of Montreal at the time, observed in his "Notes on Egyptian Antiquities presented to the Natural History Society" (1859) that "a quantity of lichen . . . placed over the front part of the body to give it a more rounded contour, and to retain the odor of the spices; and it had been swathed in numerous linen cloths, folded over the front, and with many loose pieces put in to fill out the form."

Lichens will grow just about anywhere, but not within a mummy wrapping. They need sunlight. This

Possible symbol for lichen
Hieroglyphs on Abothloé's coffin

mummy-encased lichen, despite its age, had nevertheless retained enough form that Tuckerman was able to identify it correctly as *Evernia furfuracea*, now known as *Pseudevernia furfuracea*, a European species closely related to the common antler lichen, *Pseudevernia consocians*, of eastern North America. The mummy sample also retained lichen substances that were detectable with high-performance liquid chromatography. Less sensitive thin-layer chromatography failed. No shame: it had been 2,500 years.

The source of lichens for embalming remains a bit of a puzzle. *Pseudevernia furfuracea* is not known to have grown in Egypt. Imported from less dry areas around the Mediterranean in large quantities, some of it was added to flour for the unusual, yet desirable, flavor it added to bread. Presumably, some of it was used for mummies.

P. furfuracea is still used today as a fixative in the perfume industry. Abothloé's mummy, minus a few bits of lichen, can be visited in the Redpath Museum at McGill University, Montreal, Canada.

See also perfume; Tuckerman.

Naming

Naming
Lichens have a naming problem. The Linnaean binomial system, used for all living things, is based on a unique name for each distinct living thing. It was quite reasonable that lichens were given a binomial when they were believed to be singular organisms. When Simon Schwendener upset the apple cart in 1867, disclosing the dual nature of lichens, he exposed a condition not allowed for by Linnaeus. Not only was there more than one organism present, but they were unrelated in the taxonomic hierarchy of class, order, and family. Later discoveries showed that they weren't even in the same kingdom!

The mycobiont (fungal partner) and the photobionts (algae and cyanobacteria) are distinct living things, and each has its own identity, as do the other fungi and bacteria that make up the lichen entity. The lichen had a name, and the photobionts had names, but the fungal partner had no name. In an attempt to accommodate this situation, the *Code of Botanical Nomenclature* was revised in 1950. For the sake of convenience, it stuck with the Linnaean concept of one name for one thing and decided that the existing lichen name would become the name of the fungal partner. The lichen, *as a lichen*, was left with no name. It uses the fungus name, which is an alias, a *nom de champignon*. Taxonomists these days are starting to avoid the word "lichen" when describing new species and instead are using "lichenized" or "lichen-forming fungi."

Posing another issue with lichen naming are the ongoing changes as a result of genetic analysis. Gary Lincoff, a distinguished mycologist and author of the excellent

National Audubon Society Field Guide to North American Mushrooms (1981), was often asked about writing a revised version to correct the names. He would reply, "The mushroom doesn't know we've changed its name." Thank you, Gary—the lichens, too, don't know. Given their remarkable abilities, I wouldn't be surprised if they are laughing their apothecia off and placing bets on when these large-brained creatures will figure them out.

See also Linnaeus; Schwendener.

Navigation

Tristan Gooley, a master reader of nature's signs and author of *The Natural Navigator* (2020) and *The Lost Art of Reading Nature's Signs* (2014), goes deep into the art of finding your way without the sun, a map, or a compass. One of his skills is reading lichens. Since he knows few individual lichen species by name, he relies on general characteristics, such as light preference and color changes, to make navigational decisions. For example, sunburst lichens (*Xanthoria*) have more orange compound when they grow in full sun, possibly because the orange compound screens harmful levels of UV light. When they grow in shadier spots, they are a pale orange and have a greenish tone. Those greener patches will not be on the south side of the rock (in the Northern Hemisphere). Paying very close attention to his surroundings, Gooley has noticed that the tiny apothecia on sunburst lichens are oriented toward the sun. Another example is map lichen (*Rhizocarpon geographicum*), which colonizes rocks. Lichens on the top and south-facing side of a rock will be brighter than those on the north side.

In wooded landscapes, lichens are more abundant where there's more light—perhaps because fallen trees have opened up the canopy, which would be obvious, or because you are nearing the edge of the woods. If you start seeing beard lichens (*Usnea*), you are in clean, pollutant-free air and may be getting farther away from built-up areas.

Gooley tells a story in *The Lost Art of Reading Nature's Signs* of losing his direction when the mists came down over the volcanic landscape of La Palma in the Canary Islands. "To this day I'm grateful to one particular lichen that helped me out of a challenging situation. Fortunately, earlier that day I had spotted that this grey-green lichen had a strong preference for the northwest side of the dark knobbly lava rocks, and this was all I needed to find my way down off the mountain. I never did learn its name."

Nests

Lichens are readily available for nesting material but are not as widely used as grasses, pine needles, twigs, and mud. The trouble with lichens is that many of them are so firmly attached to their substrate that birds, at least small ones, can't pick them off. Another trouble is that they can absorb a lot of moisture and lose their structure. Some birds have figured out how to use them either as components of the nest or as covering on the exterior of the nest.

Usnea species (old man's beard), which grow in great profusion in many habitats, are widely used. With their fruticose growth form, they are the same color in any orientation. The foliose lichens used for nests are the kind whose lobe tips are slightly turned up or free of the substrate. A small bird can tear those off. Unlike *Usnea*, these foliose pieces have upper and lower surfaces of different colors. To maintain any camouflage effect, the bird must place the lichen on the nest with the upper surface facing out.

Golden plovers are ground nesters. They scrape out hollows and line them with the fruticose whiteworm lichen (*Thamnolia vermicularis*), which grows on the ground in the nesting areas. The curvy white band flowing from the top of the plover's head down to its breast looks like another piece of worm-shaped lichen.

Other birds known to incorporate lichen pieces into the body of their nests are black-whiskered vireo, black-throated green warbler, cerulean warbler, Blackburnian warbler, blackpoll warbler, rusty blackbird, long-tailed tit, both evening and pine grosbeak, pine siskin, and

Golden plover

Whiteworm lichen

both the white-winged and red crossbill. Philadelphia vireos go a step further and use long stringy pieces of *Usnea* to help attach the nest to a forked twig.

Northern parulas build their nests entirely within large hanging clumps of *Usnea* lichens. The nest that they build of hair, fine grasses, and plant down is really just the lining. The body of the nest is the existing *Usnea*. They like to nest high in mature forests, where a single tree may have so many branches heavily draped with *Usnea* that a predator would not know where to start. In other parts of their range where *Usnea* is less prevalent, parulas make their nests in a similar manner using large hanging clumps of Spanish moss (*Tillandsia usneoides*) or lace lichen (*Ramalina menziesii*) instead.

In a manner similar to parula nest-building, olive-headed weavers in Madagascar and East Africa make their nests inside existing clumps of *Usnea*.

Birds that place pieces of foliose lichens on the exterior of their nests, presumably for camouflage, include eastern wood peewee, African paradise flycatcher, yellow-throated vireo, solitary vireo, red-eyed vireo, American redstart, and rufous-tailed hummingbird. Blue-gray gnatcatchers, American bushtits, and ruby-throated hummingbirds completely cover the exterior of their nests with foliose lichens.

Long-tailed silky-flycatchers in Costa Rica build their nests mainly of *Usnea*. Alexander Skutch, an author and ornithologist, observed both male and female birds bringing strands of *Usnea* to the nest site, where they expended a lot of energy kicking and tramping the lichen into a compact mass. The nest had no other lining. He also noticed that most of the nests incorporated pieces of foliose lichen. In some cases, the exterior of the nest was coated extensively in the lichen, in the manner of the ruby-throated hummingbird nest.

See also foliose; fruticose; hummingbird nest; old man's beard.

Nitrogen Fixing

The term "nitrogen fixing" suggests that there is something wrong with nitrogen. There is. The problem that needs fixing is that nitrogen bonds so tightly to itself that it's inert, yet the DNA in the nucleus of every cell—to say nothing of ATP, amino acids, proteins, and all the other stuff of life—requires it. The solution is to break the strong bond and "fix" the nitrogen to something that will give it up more readily. Nature has two ways to accomplish this. One is by lightning (violent and erratic), which produces nitric oxide and nitrogen dioxide; the other is by microorganisms, including cyanobacteria, that produce ammonia, nitrates, and nitrites.

Approximately 10 percent of lichen species contain cyanobacteria. These cyanolichens are known as nitrogen fixers, but that's just a figure of speech. Remember English class? Synecdoche is a literary device in which a part of something is used to represent the whole (as in "all hands on deck"), or when the whole is used to represent a part. Saying that cyanolichens fix nitrogen is not technically true. We are using the whole lichen to refer to its cyanobacteria; the other lichen components can't fix nitrogen. The same applies when we say that plants like peas and clover fix nitrogen. They don't. We are using the whole plant to refer to nitrogen-fixing bacteria in nodules on the roots.

Not all nitrogen-fixing bacteria are cyanobacteria. Nitrogen fixing is an anaerobic process. Underground nitrogen-fixers living inside nodules on plant roots have a naturally oxygen-free environment. Cyanobacteria, on the other hand, have to create specialized thick-walled cells (heterocysts) to exclude oxygen. One genus of the root-nodule type of nitrogen-fixing bacteria, *Frankia*, honors Albert Frank, the botanist who coined the term "symbiosis" to describe the lichen lifestyle.

Cyanolichens are important in nitrogen-poor ecosystems. Because of their captive nitrogen-fixing bacteria, they can grow where nitrogen availability is a limiting factor for plants and trees. By helping themselves, they enrich the habitat. When nitrogen compounds leach out from living lichens and dead lichens yield their nitrogen, the result is natural fertilizer, which is useful in swamps and bogs. In those environments, the normal nitrogen contribution from the decay rate of plants is slowed by high acidity levels and low oxygen in the soggy soil. A study of mesic, old-growth Douglas fir forests in Oregon and Washington indicated that more than half the annual input of nitrogen came from the extremely prolific cyanolichen *Lobaria oregana*, which grows mainly on the upper branches of trees. Ground-dwelling cyanolichens also improve the habitat wherever they grow.

See also cyanobacteria.

Nylander, William (1822–1899)

Some people are their own worst enemy. William Nylander was such a person. He advanced the field of lichenology in many and lasting ways, and at the same time he vitriolically insisted that lichens were not dual organisms.

A Finn by birth, Nylander earned a degree in medicine at the University of Helsinki and then turned his interests to biology—a common path for botanists and other biologists of the day. He became an authority on ant and bee taxonomy but soon moved on to flowering plants, bryophytes, and lichens. In the early 1850s, after he worked in the natural history museum in Paris as a traveling scholar, lichens had become his passion. Nylander wasn't happy back in Helsinki in his role as a professor of botany—he felt that the place was intellectually backward and inadequate for his purposes. He returned to Paris, where he stayed, devoting the rest of his life to lichens.

In Paris, Nylander noticed a decline in lichens and attributed it to the rise in air pollution. He thought that lichens could have a role in determining the effect of air quality on human health. He was among the first to make this connection, and he is credited with making it more widely known.

Nylander established a reputation for his classification and identification of lichens and would describe probably more species than any other lichenologist. He not only identified hundreds of species sent to him from around the world but used these specimens to build up the world's largest and richest private lichen herbarium. Housed today as an important (and separate) collection in the Finnish Museum of Natural History, it has over 51,000 species.

Nylander is also famous for pioneering the use of chemicals in identifying and classifying lichens, a process known as chemotaxonomy. It started with his observations of color changes (blue, red, or no effect) in certain parts of lichen apothecia when an aqueous solution of iodine was applied to them. He also found that drops of potassium hydroxide (KOH) and calcium hypochlorite ($Ca(ClO)_2$) produced color reactions in the presence of certain groups of lichen substances. Similar-looking lichens of different species could be separated quickly with a simple "spot test." Field guides today typically state the color to expect from spot tests for both the so-called K and C reagents for all the species for which it is useful.

His passion for lichens, which he saw as "a noble and venerable class of autonomous plants with little relationship to either fungi or algae," ran deep. Schwendener's theory of lichens as dual algo-fungal organisms was "heresy" to Nylander, and he took it personally; he saw the theory as "absurd," an "absolute fantasy," and even as a "calumny." Although his misguided passion, disparaging remarks, and solipsism alienated other lichenologists, Nylander remains an important figure in the history of lichenology.

See also lichen substances; Schwendener.

Old Man's Beard

Lichens in the genus *Usnea* are called "old man's beard," or sometimes just "beard lichens." They are yellowy-green to gray-green and threadlike. Some are stringy and dangle in pendulous tangles, possibly like an old man's uncombed beard. Others are more shrubby and form upright clusters. *Usnea* species are easy lichens for beginners to identify to genus. The strands are more or less circular in cross-section, and inside each strand is a central cord. The cord has some elasticity; it's not bungee quality, but you can confirm it with a gentle pull. Other filamentous lichens don't have a cord.

These fruticose lichens grow on trees and shrubs and rarely on rock. They can be found on every continent, including Antarctica. It is a huge genus, with over 300 species, and identifying them beyond genus can be a challenge, even for the experienced.

Old man's beard likes the light. It grows well in the canopy, and you can find it on twigs and branches on the ground after a storm. You can also find it on shrubs around pond edges where at least one side of the shrub is open to the sun. It colonizes the twigs of dead trees no longer shaded by leaves, causing people to wonder if the lichen killed the tree. It didn't.

Different species of old man's beard have been used all around the world for a variety of medicinal purposes, including lung troubles, stomachache, ulcers, altitude sickness, and uterine problems. The bioactive compound is usnic acid. For external use, it is powdered and applied as a paste or poultice to wounds, athlete's foot, boils, and blisters. The most common use today is as an antibacterial and antifungal agent in both home-crafted and commercial products.

Because of its stringy nature, old man's beard has been used as a fiber for stuffing pillows, cushions, mattresses, and all manner of bedding material, and it has even been plaited into an emergency backwoods blanket. I guess that could work well, but only if kept dry because it is also known for its absorbency. Old man's beard has been used for baby diapers, in feminine

There was an Old Man with a beard, who said, "It is just as I feared!—
Two Owls and a Hen, four Larks and a Wren,
Have all built their nests in my beard!"

hygiene, and packed into nostrils to stem a nosebleed. Not being fond of scratchiness, I personally would save these uses for last resort.

Old man's beard, along with some other lichens, is useful to humans in two other ways; it is especially sensitive to sulfur dioxide pollution and serves as a bio-indicator of air quality, and it is a source of dye.

Old man's beard is useful to birds as well, being used as nesting material by several species. Maybe Edward Lear had that in mind when he wrote his famous limerick about an old man with questionable hygiene.

See also fruticose; nests.

Orchil

Purple dye has been on a long, strange trip. It started with murex dye from Mediterranean shellfish, split into parallel paths with lichen dyes (orchil and cudbear), and ended up being synthesized. As far as we know, humans "discovered" the color purple between 16,000 and 25,000 years ago in France by mixing hematite and manganese for use in their cave paintings. The first purple dye we know of was discovered in Tyre, Lebanon, around 1575 to 1200 BCE. One of the myths, believable by all dog owners, is that the dog of a shepherdess got a purple stain on its snout after tearing apart a shellfish on the beach. If not this misadventure, something led to the discovery that certain shellfish produce a sticky fluid that oxidizes in air and becomes a brilliant purple. It was a permanent dye that became known as Tyrian purple, or murex, for the shellfish. The problem was that it was outrageously expensive, since it took thousands of shellfish and a complicated process to garner a single gram of dye. Consequently, a purple garment was a symbol of wealth and power. The motivation to find another source would have been high, and going higher as the shellfish were depleted, when somehow someone discovered that certain lichens could produce a purple dye.

Roccella lichens, found on rocks on the shores of the Mediterranean, yielded a purple dye that came to be known as "orchil" (sometimes "archil"). It had the advantage of being relatively inexpensive. The process involved breaking up the lichen, steeping the pieces in urine (as a source of ammonia), stirring it frequently for aeration, and waiting approximately three weeks for the purple color to develop. Dyeing with orchil before overdyeing with murex reduced the amount of murex needed. Not only did this process reduce the price, but the resulting color was less susceptible to fading in bright light than purple from orchil used on its own.

After the fall of the Roman empire, not much is written about orchil until Federigo, a merchant of Florence, returned from the Crusades with the recipe. He built a hugely successful, monopolistic dye business that lasted several generations. Pietro Antonio Micheli, the eighteenth-century botanist, wrote in *Nova Plantarum Generum* that the dye lichen was known informally as roccella, orcella, or raspa, and that the dye itself was known as oricella. In a history of Federigo's family, an unknown descendant wrote that the lichen "gave so much wealth and good living to our family" that the family adopted the name "Orcellai," which morphed into "Rucellai" and became synonymous with purple. The lichen genus *Roccella* is named for Federigo's family.

The ammonia extraction process with *Roccella* lichens converts the depside erythrin into orcinol, which is then converted to orcein. It turned out that other quite different lichen species with different depsides also yielded orcinol and orcein if subjected to the ammonia process. The resulting dyes were not called orchil (some were called cudbear), but they were orchil equivalents. Evidence from burial sites in Denmark and Greenland shows that orchil equivalents were used in other parts of the world as early as the Bronze and Iron Ages. What was the common thread linking different lichens, different cultures, different countries, and different habitats? Urine?

See also cudbear; dyes.

Parton, Dolly (1946–)

Jessica Allen, Associate Professor of Biology at Eastern Washington University, and her fellow lichenologist James Lendemer, Staff Lichenologist and Associate Curator of the Institute of Systematic Botany,

New York Botanical Garden, discovered a new species of lichen on Hangover Mountain in the Unicoi Mountains of North Carolina, near the Tennessee border. Technically speaking, this green crustose lichen had been found before, but without the fruiting bodies (apothecia) that were necessary to determine its full identity. This new find had apothecia, making it possible to confirm that it was indeed a new species. Naming it was their privilege. It belonged in the genus *Japewiella*, but the species epithet (the second part of the binomial) was up to them. They chose to call it *Japewiella dollypartoniana*, and less formally "Dolly's Dots," because the brown apothecia looked like dots on the green thallus.

The name is anything but frivolous. Allen and Lendemer took several factors into consideration in choosing it: the lichen grows in the southern Appalachians not far from where Dolly Parton was born; it was a way to honor her contributions as a country music legend and a philanthropist; it began to compensate for the underrepresentation of women in the naming of species; and not least, it engaged people far beyond the boundaries of lichen academia. Any news about Dolly Parton is news. It's quite possible that some Parton fans became aware of lichens for the first time in their lives. There's a country song waiting to be born, about hanging out on Hangover Mountain, where my headache vanished and my love for lichens took its place. Dolly?

See also careers.

Peppered Moth

Lichen mimicry is a useful camouflage strategy as long as the creatures adopting it can count on the lichens. This is not usually a problem because lichens are long-lived, stationary, and ubiquitous, but the Industrial Revolution changed that. Most lichen species are susceptible to air pollution, and the massive change in air quality caused by industrialization not only killed lichens but also darkened the substrate with soot. Even without any sooty deposits, tree trunks and branches showed their natural darker color when the paler lichens on them were gone.

The peppered moth (aka pepper-and-salt moth) got its name from its white body and wings speckled with black. This mottled appearance gives it excellent daytime camouflage on the lichen-covered tree trunks where it rests. As with lots of pale-colored critters, genetic mutation in peppered moths sometimes results in dark coloration, a condition known as melanism. Before the Industrial Revolution, the melanistic population was small, since dark ones were more easily spotted by predators and killed. In the nineteenth century in cities in the United Kingdom, people started to notice that the melanistic form of peppered moths was becoming more common than the speckled form. Lichen mimicry was no longer successful, as the pale speckled moths were now the ones being picked off by predators.

The moth life cycle is short, so the change in populations from pale moths to dark ones happened relatively quickly. In the heavily industrial city of Manchester, England, the first dark peppered moth was noted in 1848, and less than fifty years later 98 percent of the moths in the city were the dark ones. By the mid-twentieth century, air-quality legislation had led to less soot, less sulfur dioxide, the return of lichens, and an increase in the population of paler peppered moths.

The moth is also called Darwin's moth because it so vividly illustrates natural selection.

See also air pollution; camouflage.

Perfume

Lichens have played a role in the fixing of perfumes since at least the twelfth century. Saladin (1137–1193) mentioned this usage in his *Compendium Aromatariorum*. In much earlier times, lichens were used in Egypt as part of the mummification process. Compounds in the lichens "fix" volatile perfumes to the skin to keep them from evaporating too quickly. Two lichens with compounds that work as fixatives are *mousse de chene* in French ("oak moss" in English, *Evernia prunastri* in Latin) and *mousse d'abre* ("tree moss," *Pseudevernia furfuracea*). The French names are used outside of France, perhaps because of the influence of the perfume trade. Neither species is a moss. Both have the shrubby growth form of fruticose lichens. Oak moss grows mostly on the bark of hardwoods and tree moss mostly on evergreens.

The perfume industry uses tons of these lichens annually. Most are collected in Europe and North Africa. A friend remembers harvesting lichen in the late 1970s near Millau in the Occitania region of France during a semester abroad. She was working on a goat farm, where lichen gathering in the nearby woods was an occasional side task. The lichen was not for goat fodder but for the perfume industry. The sample she saved was oak moss, *Evernia prunastri*.

The actual perfume fixative is a blend of compounds extracted from the lichen with organic solvents like alcohol, benzene, acetone, or ethylene glycol in a complex process. The final extract has a mossy scent of its own and is most compatible with ferny, new-mown-hay types of perfumes. Oak moss extract is a better fixative and has a better scent than tree moss. Too bad that the yield is higher from tree moss. The industry responded by blending them.

Although oak moss and a tree moss closely related to the European one grow in the United States, they fortunately are not used in the perfume business. Commercial use of "wild" lichens is not sustainable, and so far we don't have a way to farm them.

See also mummification.

Perithecium (*plural* Perithecia)

Perithecia are one of the two main types of fruiting bodies of all lichens (except the basidiolichens) that make spores from sexual reproduction of the fungal partner. (The other type is apothecia.) The word comes from the Latin prefix *peri-*, which means "enclosing" or "surrounding" (as in "pericardium," the membrane around the heart) and the Greek *thēkion*, meaning "small case" or "box." Despite the etymology, the small case of a perithecium is usually described as a flask. Think of a rounded glass flask in a chemistry lab versus the flattened hip flask in your grandfather's pocket. The flask is more or less enclosed within the lichen thallus. Its presence is detected by the opening through which the spores are released or sometimes by the neck of the flask if it protrudes. Apothecia, on the other hand, are highly visible on the surface of the thallus or raised above it. If apothecia are the extroverts of lichen fruiting bodies, perithecia are the introverts.

As with apothecia, the spores are produced in sacs, called "asci" (singular, "ascus"), which are clustered on the inside walls of the perithecium.

On the off chance that you want another lichen glossary word, I offer "pyrenocarp." It's a synonym for "perithecium." The rationale is that the fruiting body can be seen as pear-shaped versus flask-shaped, and *Pyrus* is the genus for the common pear. Taking this further, a genus of lichens called *Pyrenula*, with forty-eight species in North America, has perithecia for fruiting bodies. The names can be helpful or confusing. It depends on whether your flask is half full or half empty.

See also ascomycota; apothecium; basidiolichens.

Pliny the Elder (23–79 ce)

Caius Plinius Secundus, aka Plinius, aka Pliny the Elder, was a military commander in the army and navy of the Roman empire, but he is best known as a prolific writer. A workaholic, he wrote at night after his imperial duties were done for the day. Of the more than 200 books he wrote, only the 37 books that make up *Natural History* survive today. The section on the medicinal use of plants included prescriptions using lichens.

Roughly 300 years after Theophrastus first used the word "lichen" to apply to lichens, the word turned up again in the written record—this time in Pliny's *Natural History*, which became a model for encyclopedias. Although not arranged alphabetically, it covered multiple, extremely diverse topics. It was an aggregation of existing knowledge, bad science, and myth. There are accounts of headless people and dog-headed people who communicate by barking. Did he not have a fact-checker?

Books 20–29 of *Natural History* addressed medicine and drugs from plants. Pliny drew heavily from Theophrastus. He listed over 900 drugs, compared to 600 cited by Dioscorides (his contemporary) and 550 by Theophrastus. Plant taxonomy of the day was rudimentary. Lichens and mosses were not clearly differentiated. Pliny prescribed the use of lichens for the treatment of skin diseases and other ailments. In book 26, after naming several plants for the treatment of tetter (skin diseases, including lichen planus), he writes: "Lichen is the herb preferred to all . . . It grows among rocks, and has a single broad leaf near the root, a single long stem, with long leaves hanging down." It's dubious that he is describing what we know today as a lichen, but he goes on to say: "There is another kind of lichen also adhering to the rocks so much, like moss, which is applied. Dropt into wounds, or applied to abscesses, has the property of arresting haemorrhage. Mixed with honey, it is curative of jaundice, the face and tongue being rubbed with it."

In book 27, Pliny writes of a moss, which some translators think is actually a lichen. "There [it] grows near running streams, a dry, white moss, upon ordinary stones. One of these stones, with the addition of human saliva, is rubbed against another; after which the first stone is used for touching impetigo, the party so doing uttering these words: *Cantharides begone, a wild wolf seeks your blood*." (Cantharides is a type of blister beetle, also known as Spanish fly, whose secretions cause blisters and reddened, burning skin.) The efficacy of this incantation is dubious. It may have needed the help of a full moon.

Although there are no lichen substances known to cure impetigo or jaundice, there are several with anti-bacterial properties that could be helpful in the treatment of wounds and broken skin.

Pliny the Elder survived the horrific reigns of Caligula and Nero. He died during the reign of Vespasian while attempting to rescue people the day after the eruption of Mount Vesuvius. His legacy is *Natural History* for two reasons. First, it is a detailed record (over one million words) of the knowledge and beliefs of the first-century Romans. Even the sensational and inaccurate entries have value as a view into the times. (Which of our views will seem ludicrous in 2,000 years?) Second, it is the original prototype of an encyclopedia—the broadest possible collection of the knowledge of the day to be used as a resource. *Lichenpedia* is the tiniest of tributes to Pliny, the father of the encyclopedia.

See also dermatology; Theophrastus.

Poetry

Lichens can't fail to inspire poets—once they notice them. Lew Welch (1926–1971), Beat Generation poet and travel companion to Jack Kerouac, honored these tiny things with modest needs that break down rock in his poem "Springtime in the Rockies, Lichen."

All these years I overlooked them in the
racket of the rest, this
symbiotic splash of plant and fungus feeding
on rock, on sun, a little moisture, air—
tiny acid-factories dissolving
salt[s] from living rocks and
eating them.

Jane Hirshfield (1953–), author of nine poetry collections, called lichens "chemists of the air" in her poem
"For the Lobaria, Usnea, Witches Hair, Map Lichen,
Beard Lichen, Ground Lichen, Shield Lichen." She captured the mystery and sense of age felt by many when
they think of lichens.

When I saw you, later, seaweed reefed in the air,
you were grey-green, incomprehensible, old.
What you clung to, hung from: old.

Lichen patterns
on rocks

John Ellor Taylor (1837–1895), a Victorian science writer, appreciated lichen patterns on rock. In "Moor and Mountain," he expressed it in a heroic couplet, his favorite poetic form: "Art's finest pencil could but rudely mock / The rich grey lichens broider'd on a rock."

Sarah Lindsay (1958–) writes of a lichen in her perfectly titled poem "Stubbornly" (2008): ". . . growing slightly slower than stone . . . etching on the unmoved rock / the only rune it knows."

My favorite, because of its sense of place and the wonder of our connectedness with all things, is Lawrence Millman's prose poem "Lichen," which opens: "May the gods of the tundra grant me lichen until I become lichen myself."

See also Thoreau.

Poikilohydry

"Poikilohydry," the term for the inability of an organism to regulate its water content, is pronounced poy-kill-oh-high-dree. These organisms are at the mercy of the amount of moisture in their surroundings. The condition is analogous to cold-bloodedness in animals that lack the ability to regulate their body temperature and are at the mercy of the amount of heat in their surroundings. Whether it's water or warmth, both have to adapt.

The best-known poikilohydric organisms are mosses among plants, tardigrades among animals, and lichens among fungi. One of the necessary coping mechanisms is desiccation tolerance. Tardigrades take the gold medal in that event and in many other stress conditions. They are the winners in the extremophile decathlon, but lichens are also survival experts. When they're dry, they exhibit little to no physiological activity—they slow down and wait. They photosynthesize when they can—in rainy, dewy, or humid weather. Periods of dehydration may be short or long. Wet-dry cycles might occur daily for some desert-dwelling lichens, owing to the frequency of dewfall. Other species manage to cope with an irregular wet-dry cycle. Some can revive fully after years of desiccation. Very few lichen species are aquatic. Most are far better able to handle too little water than too much.

Many foliose and fruticose lichens become brittle when dry. In that state, fragments are easily dislodged and dispersed. If the pieces dislodged from an original spot that was too dry for too long land elsewhere in a suitable place, they may be given a chance to start new colonies. Some crustose species break up into little chunks (areoles) and look like the cracked surface of sun-dried mud. Each chunk is thought to be functionally independent, but they will reunite into a single entity with a continuous surface when wet.

Official lichen descriptions are based on the characteristics of dry specimens. Photographs in field guides are always of the dry lichen—except when photographers can't help themselves and publish photos of the hydrated specimens because they are so beautiful. These photos are usually annotated to reduce confusion.

See also BIOMEX; tardigrades.

Quandaries

Lichens have always been mysterious. Their symbiotic lifestyle as a tightly coupled alga and fungus has been known since 1867. It is also known that the alga is sometimes accompanied by or replaced by a cyanobacterium. Around 2010 we learned that other fungi are present, not just those sometimes growing *on* the surface (lichenicolous fungi) but others *inside* the lichen (endolichenic fungi). How many species are there, and what are they doing? Do they have a role related to the more than 1,000 chemicals that lichens make? Have lichens been so difficult to cultivate because these other fungi were absent? Are they necessary for the creation of a lichen thallus?

Consider the specific quandary that was plaguing Trevor Goward in British Columbia, Canada. Goward is the best kind of naturalist: highly self-educated, with no biology or science degrees, he works in the field, where he looks and he thinks. He knew that black tree-hair lichen, *Bryoria fremontii*, is eaten by indigenous peoples, and that high levels of vulpinic acid make yellow horse-hair lichen, *Bryoria tortuosa*, poisonous. He knew the subtle differences in the morphology and habitat of these two lichens. He also knew, from genetic analysis done in 2009, that they share the same fungus and the same alga. What?! That should define them as the same species. Goward wondered if a third party, perhaps bacteria, made the two lichens different.

The conundrum bothered Goward. He took it to Toby Spribille, a young lichenologist who knew first-hand the bias against the self-educated. Spribille had spent his childhood in northwestern Montana being home-schooled, which isn't helpful for admission to most US universities. Determined, he took and passed the high school equivalency test, got accepted into college in Germany, and eventually earned a PhD in

Austria. He and Goward knew each other and were well matched. He accepted the challenge of Goward's puzzle. After much genetic analysis, he discovered another fungus, a single-celled basidiomycete yeast in the cortex of both *Bryoria* species. It was not a contaminant; it was necessary for each lichen to be what it was. He then confirmed that related yeasts were present in many different lichen genera from six continents.

In 2016, Spribille and his colleagues published their findings, which documented for the first time a possible role for an additional fungus. The basidiomycete yeast was much more abundant in the yellow *Bryoria*, suggesting that it is implicated in the production of vulpinic acid. If so, how and why? Also, when trying to cultivate lichens from their fungal and algal components, one of the problems is that the cortex doesn't fully form. Does this cortex-dwelling yeast have a role? How limited is our understanding of symbiosis? Can we stick with the current method of naming lichens?

So many questions. Spribille is now Assistant Professor, Department of Biological Sciences, University of Alberta, Canada, with a lifetime of quandaries ahead of him.

See also black tree-hair; naming.

Reproduction

Lichens are composed of more than one organism, so reproduction of the whole entity is a bit of a challenge. How do the components stay together? Do they need to? If they reproduce separately, will they find suitable partners again? Part of the solution is simple fragmentation. Pieces of thallus break off, blow around, and if they happen to lodge where light, moisture, and substrate are suitable, they will grow back into a whole lichen. The components needed to form a new lichen are all there.

More important to reproduction are special reproductive structures that also contain all the components necessary for a new lichen. These structures, or propagules, mainly soredia and isidia, are designed to be easily dislodged and dispersed. These methods are vegetative and do not increase genetic diversity. It's like growing a new plant from a cutting.

Sexual reproduction of a lichen as a complete (dual) entity makes no sense. The photobiont is usually completely enclosed within the thallus and reproduces in place, asexually, by simple cell division. Only the mycobiont (fungal component) reproduces sexually. After fertilization, it forms a fruiting body where spores are created and then released. The germinating spore ventures out, like a hunter-gatherer, to find a suitable alga (or cyanobacterium) to capture. Suitability is determined by a chemical "lock-and-key" mechanism. The developing fungus secretes a protein with a particular three-dimensional shape (the lock). As the fungus grows, it needs to encounter an alga of the right shape (the key) to fit its lock, or else it dies.

There are exceptions. Some fungi associate with different species of algae in different parts of their geographic range. It's as if some algae have a master key. *Trebouxia*, for example, form lichens with many different fungi. A few fungi have no need for the hunter-gatherer lifestyle. Algal cells reproduce within the fungal fruiting body, adhere to the spores, and stick with them on dispersal. An example is the *Staurothele* genus, aka rock pimples. "Whither thou goest, I will go."

Many lichens have their virtual eggs in multiple baskets—they reproduce sexually by spores and asexually by propagules. Although capable of both methods, many heavily favor asexual reproduction. For example, both the common greenshield lichen, *Flavoparmelia caperata*, and monk's hood lichen, *Hypogymnia physodes*, rely on soredia. Both can produce fruiting bodies but seldom do. Another example is lungwort, *Lobaria pulmonaria*, though this lichen has hedged its bets on the asexual side by having both soredia and isidia.

On the other hand, lots of species do have their eggs in one basket. Some have never been known to produce a fruiting body. For example, dust lichens (*Lepraria*) have devoted their entire thalli to dispersible granules—no sex at all, ever. Christmas lichen (*Herpothallon rubracincta*) generates red granules (for dispersal) on the older parts of its thallus, and no one has ever seen fruiting bodies. Other lichens take the opposite approach and reproduce only sexually. Still others produce both sexual and asexual spores. It takes all types.

See also dispersal; isidium; lichenicolous fungi; soredium.

Rhizine

"Rhizine" is pronounced to rhyme with "mine" and "thine," or "rise and shine." The name comes from the Greek root *rhiza*, meaning "root." Just as rhizomes in plants are not roots but underground stems that may be mistaken for roots, so rhizines in lichens are also not roots. They are fungal filaments on the lower surface of

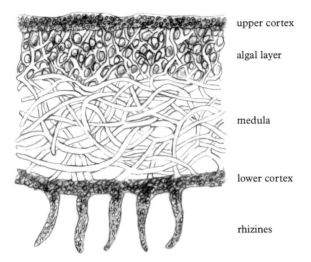

upper cortex

algal layer

medula

lower cortex

rhizines

foliose lichens. Their sole job is to connect the lichen to its substrate. They belong to the Bill Belichick school of football coaching: "Do your job." They do not absorb water, transport nutrients, produce spores, or kick field goals. They hold the lichen in place. That is their job, at least from the perspective of the lichen. For humans, they have another "job."

Not all foliose lichens have rhizines. (There are other methods for anchoring the thallus, and there are even a small number of so-called vagabond lichens that wander.) The presence of rhizines on a lichen provides us with clues to help identify it. For example, rhizines on a given species of lichen may be black or white, dense or sparse, scattered or clustered. Another clue to species identification comes from the nature and degree of branching on an individual rhizine. Some are simple and straight, like a tap root on a plant. Others are branched, and the overachievers have branches on the branches. Other filamentous lichen structures called "cilia" can be confused with rhizines at first glance because they also appear to be coming from the lower surface. Like rhizines, cilia are made of fungal threads, but closer inspection shows that cilia are confined to the edges of the thallus and their tips are not fastened to anything.

See also foliose; substrates; thallus.

Rock Climbing

Steep cliff faces have ledges and crevices that make footholds for rock climbers and lichens. Neither wants to fall. Are they becoming mutualistically symbiotic? Climbers are gaining appreciation for what they share the rock face with, and lichens are being considered when choosing climbing routes and establishing new routes. Climbers are discovering a new worldview, and lichens are getting habitat protection.

Climbers and lichenologists make awesome teams who can help resolve the ever-present conflict between recreation and nature. Together they discover what lichens grow in these relatively unexplored, specialized habitats and answer the questions of land managers and conservationists about the effect of rock climbing on cliffside ecology. The "glass half empty" answer is that climbers inadvertently dislodge lichens. The most vulnerable sites are those with the highest climbing use intensity—in other words, highly rated climbs with convenient parking. The most vulnerable lichens are foliose species with weak attachment. The "glass half full" answer is that awareness leads to protection.

Peter Clark, Forest Ecologist at the University of Vermont and an experienced rock climber with degrees in forest ecology and biogeography, examined cliffside vegetation in a popular climbing destination, New River Gorge in West Virginia, on behalf of the National Park Service. He has also climbed cliffs across the country and in Canada with an eye to documenting species abundance and diversity and assessing climber effects. He rediscovered frosted rock tripe (*Umbilicaria americana*) in West Virginia, where it was thought to have been lost. He also found cliff gold dust lichen (*Chrysothrix susquehannensis*), which was previously known from only a single site in Pennsylvania. Clark found it at fourteen more sites, usually on exposed cliff faces at heights above eighty feet. His work, along with that of other climbers and lichenologists, has led not only to recommendations for land managers and climbers but also to new information on lichen biogeography.

Climbers who are aware of lichens can't help but respect them. Both are extremists with skills to survive in challenging places.

Rock Tripe

If you see a large-lobed foliose lichen growing on a siliceous rock, such as granite, it could well be from one of a group of lichens known as rock tripe (aka *tripe de roche*). Some species in the genus *Umbilicaria* have common names describing their appearance, such as fringed rock tripe, smooth rock tripe, and toadskin lichen. These rock tripes are attached to the rock at a single point by a cord of fungal fibers reminiscent of an umbilicus, the source of the genus name *Umbilicaria*.

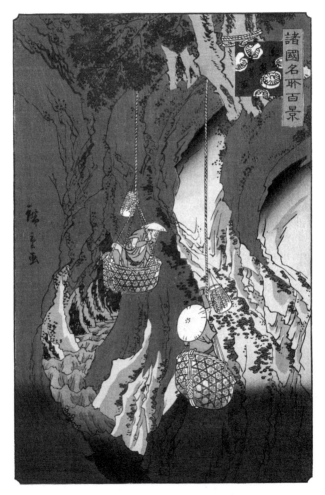

Gathering iwatake *Umbilicaria esculenta*
Courtesy of the Library of Congress, LC-DIG-jpd-01373

Many of the rock tripes grow and thrive year-round in the harsh climate of arctic latitudes and can be a source of food when there is little else. Inuit peoples and stranded explorers eat them as a food of last resort. Others eat them by choice. Palatability varies with species. The Cree would add *Umbilicaria muehlenbergii*, the plated rock tripe, to their fish broth as a thickener and for extra nutrition. *Umbilicaria esculenta*, known in Japan as iwatake, is so desirable that people used to risk rappelling down steep cliffs in wet weather to harvest it. A woodblock housed at the Library of Congress shows men in baskets being lowered down a rock cliff to gather it. *Esculenta* is Latin for "edible" or "fit to be eaten" and is used in many plant and fungi names. For example, one of the highly desirable morel mushrooms is *Morchella esculenta*, and the lichen thought to be the manna of the Israelites is *Circinaria esculenta*.

Rock tripes are among the many lichens that are used as a source of dye. The lichens themselves are not brightly colored, but an ammonia extraction process yields purple, reds, and pinks.

See also dyes; food (for humans); Franklin's expedition; survival food.

Ruskin, John (1819–1900)

Artist, author, philosopher, teacher, conservationist, and social revolutionary are all roles that apply to John Ruskin, but he is best known as an influential art critic. He was a polymath whose interests extended to architecture, geology, ornithology, botany—and lichens. His abiding fascination with lichens went beyond their interesting form, their variety of colors, and their place in art.

It was their message of endurance and their relationship to environmental change that intrigued him.

Ruskin advocated paying very close attention to detail in nature. In *The Elements of Drawing* (1856–1859), he wrote: "The grass may be ragged and stiff, or tender and flowing; sunburnt and sheep-bitten, or rank and languid; fresh or dry; lustrous or dull: look at it, and try to draw it as it is, and don't think how somebody 'told you to do grass.' Or how to 'do a stone.'" Grass and stones are familiar objects. Lichens were not—and are not. Ruskin used them for exercises in the kind of direct observation that he believed was important in art and life. He didn't want his art students to attempt to draw landscapes until they had demonstrated their ability to draw the lichens on a stone. "Be resolved, in the first place, to draw a piece of rounded rock, with its variegated lichens, quite rightly, getting its complete roundings, and all the patterns of the lichen in true local color. Till you can do this, it is of no use your thinking of sketching among hills; but when once you have done this, the forms of distant hills will be comparatively easy."

The ubiquitous presence of lichens on rocks and trees could not help but contribute to the look of a landscape. "Always draw whatever the background happens to be," Ruskin advised, "exactly as you see it." He used background lichens as a litmus test when critiquing art. "The wild sea-weeds and crimson lichens drifted and crawled with their thousand colours and fine branches" on the Venice canals, but not in Canaletto's paintings, which he deemed mechanical.

Ruskin admired the work of the artist J.M.W. Turner for the most part, but had to point out how in one painting he had "absolutely stripped the [projecting] rock of its beautiful lichens to bare slate." Familiar with the scene, he knew the rock to be "covered with lichen having as many colours as a painted window." Ruskin praised a work titled *The Moorland* by J. W. Inchbold, now in the Tate Gallery in London, as "more exquisite in its finish of lichenous rock painting than any work I have ever seen."

He deplored the then-innovative style of James Mac-Neill Whistler's paintings. Whistler sued him for libel over a scathing dismissal of his *Nocturne in Black and Gold: The Falling Rocket*. In this case, it was not a lack of lichens that bothered Ruskin, but a general lack of "truth to nature." Whistler won the case, but so did Ruskin (in a way): damages were assessed at one farthing—a quarter of a penny.

Ruskin's commitment to detailed observation in nature was shared by the modern Pulitzer Prize–winning poet and essayist Mary Oliver, who declared in her poem "Yes! No!": "To pay attention. This is our endless and proper work." She also wrote, perhaps pleaded, in "Sometimes": "*Pay attention. / Be astonished. / Tell about it.*" In a similar way, Ruskin encouraged artists to "tell about it"—accurately, visually—and to not forget the lichens.

Schwendener, Simon (1829–1919)

At the annual meeting of the Swiss Natural History Society in the fall of 1867, Simon Schwendener opened a can of worms. He launched the botanical session with a topic that was soon described as a "notorious hypothesis." He presented his theory that the two components of a lichen, green cells and white fibers, are not genetically related. This was heresy. A lichen was one thing, a singular organism, a plant—wasn't it? One of his former professors who had been anxious to hear from his talented student was very skeptical and wrote: "According to the current view of the speaker, lichens have to be seen not as autonomous plants, but as fungi in connection with algae." I sense the invisible exclamation points. William Nylander, a respected Finnish lichenologist, dismissed the idea with as much condescension as he could muster, accusing Schwendener of "being led astray by inexperience." Another lichen expert, Lauder Lindsay, a Scot, found the idea to be "merely the most recent instance of German transcendentalism applied to lichens."

The hypothesis was eventually proven to be true, but it was not until the start of the twentieth century that it was widely accepted that lichen is not a singular organism. As with many scientific breakthroughs, Schwendener hadn't arrived at the idea on his own. Anton de Bary, a German mycologist, had already discovered that *Nostoc* (a type of cyanobacteria) would form an actual lichen when associating with certain fungi. He saw that the lichen was not a singular organism. Schwendener built on the idea, supported it with extensive microscope work, announced it to the scientific community, took the heat, and continued to substantiate it, supported by some open-minded scientists. Albert Frank, a German botanist, put forward the term "symbiosis" to encapsulate the fungal-algal relationship. De Bary adopted the term, popularized it, explored the idea of "living together," and extended it beyond the mutualistic nature of lichens.

Schwendener, the only son of a farmer, was expected to take over the family farm, but his heart wasn't in it. He was an excellent student with a growing interest in science. With limited resources, he managed to enroll in classes at the University of Zurich, where he eventually attracted the attention of people who could mentor him and support his advancement. He was a skilled user of the microscope and became one of the best of his time.

It was about 200 years earlier that another highly skilled microscope user had also had his findings rejected. Antonie van Leeuwenhoek, known as the father of microbiology, started as a draper and had no formal scientific training. He learned to grind lenses and then made his own microscopes, which had unmatched clarity and resolution. He was the first to use the microscope for discovery and not just for enlarging views of known things. Others with poorer microscopes could not see his animalcules and couldn't or wouldn't believe his discoveries. Both Schwendener and Leeuwenhoek had the "wrong" backgrounds for their findings to be accepted by the establishment.

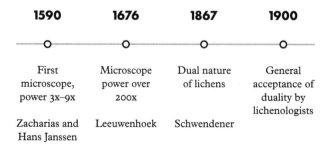

1590	1676	1867	1900
First microscope, power 3x–9x	Microscope power over 200x	Dual nature of lichens	General acceptance of duality by lichenologists
Zacharias and Hans Janssen	Leeuwenhoek	Schwendener	

Schwendener finally gained recognition and, after several posts, was offered a chair at the University of Berlin, where he was ahead of his time in recognizing women's rights. He recommended for membership three of the only five female members of the German Botanical Society in its first fifteen years. He also had at least one female postdoc student. Other science professors did not permit women to attend their lectures. If one was spotted in the lecture hall, the "respected Miss" was escorted out.

Simon Schwendener is credited with the discovery of the dual nature of lichens, but Anton de Bary had led the way. The thought that lichens might not exist autonomously, combined with the new term "symbiosis," triggered research into the phenomenon of symbiosis in other fields.

Smith, Annie Lorrain (1854–1937)

The British lichenologist Annie Lorrain Smith was one of the women accepted into the Linnean Society in 1904 when it revised its charter to allow women to become "fellows." The organization, founded in 1788, is the world's oldest active natural history society and has a tradition of being apolitical and nonsectarian. Its motto is "Naturae discere mores," meaning "To learn the ways of nature"; until 1904 the unspoken part of the motto was ". . . as long as you are male."

Smith studied at the University of Edinburgh but vehement opposition to higher education for females prevented her from graduating. She studied French and German abroad, came back to London, and was able to take classes at the Royal College of Science. Her teacher, D. H. Scott, found a job for her at the British Museum; she would work there for forty years but was never, being female, on the payroll. Scott made arrangements for her to be paid through a third party. During her time at the museum Smith described over 200 new taxa. She was the team leader of a natural history survey in Ireland that included credentialed scientists from Europe. Ten years later, in 1921, she published *A Handbook of the British Lichens* with keys to all species and also *Lichens*, which remained in use as a standard text for about fifty years.

Smith twice served as president of the British Mycological Society, of which she was a founding member. In 1931 she was granted a pension by the Crown for her "services to botanical science" and further recognized in 1934 in the British honors system with an OBE (the Most Excellent Order of the British Empire) for her "contributions to mycology and lichenology."

Beatrix Potter, a contemporary of Annie Smith's and a serious mycologist and lichenologist herself, advanced the knowledge of fungal-spore development. Both of them had to have men read their papers to the Linnean Society in their absence. Potter's work was not supported, and she is best known today as an author and illustrator of children's books, including *The Tale of Peter Rabbit*. She was thirty-eight years old when Annie Smith and fourteen other women (not including her) were admitted into the Fellowship of the Linnean Society. Its official website today claims that "we are committed to increasing the diversity of our Fellowship, and to providing opportunities for those usually excluded from the study and appreciation of natural history." It also acknowledges that "Potter was a capable and enthusiastic mycologist . . . [whose] strengths lay in her meticulous observation and artistic prowess." Discerning any condescension in the tone is up to the reader. The executive secretary of the society noted in 1997 that Potter had been "treated scurvily." About 100 years overdue. Scurvy treatment was not restricted to females: when Darwin's paper on the theory of evolution had its first presentation at the Linnean Society, it got the silent treatment, owing perhaps to the "subject being too novel and ominous." Darwin prevailed.

Annie Lorrain Smith remains an obscure lichenologist, like so many of them, and the marine lichen *Verrucaria lorrain-smithiae*, named in her honor, is equally obscure.

Soredium (*plural* Soredia)

Soredia are little lichen bundles that protrude through breaks in the lichen cortex. The singular form of the word, soredium, is a technical term only: you will never see a soredium on its own because soredia are not solitary things—they cluster. Each soredium (a tiny fraction of a millimeter) is made up of a few fungal threads wrapped around a few algal cells. They are easily dislodged—by wind, by animals, by rain—because their purpose is dispersal. Sexual reproduction (by spores) is limited to the fungus, whose germinating spore needs to find a suitable photobiont (algae or cyanobacteria) to capture and form a lichen. Soredia don't have that problem. They already contain the necessary components to start a new lichen, a clone of its "parent"; they just need to land in a place where they can get established. Think

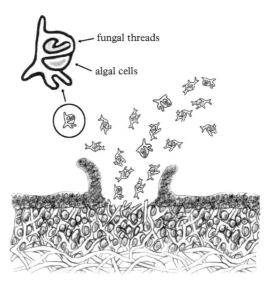

of a gardener taking a cutting to start a new plant. The cutting has the advantage of the gardener's careful choice of a medium or habitat where it is likely to grow. Soredia make up in volume what they lack in choice.

Not all lichens produce soredia, so their presence or absence is diagnostic for identification. So too is their location on the thallus. Soralia, the structures where soredia are formed, have different shapes—round, oval, elongated, and crescent-shaped. They may be confined to the edges of the lobes on a foliose lichen (giving it a frilly look) or clustered on the tip of a lobe, the underside of the upturned tip of a lobe, or the upper surface of the thallus. In some cases they cover the entire thallus, replacing the cortex and giving the lichen a "leprose" appearance. The soredia may be fine and powdery or coarse and granular.

Marcello Malpighi (1628–1694), an Italian surgeon and biologist, was the first person to describe soredia. He realized that their role was for propagation, which is why he called them "seeds." This Malpighi is the same person whose name is carried forward in human anatomical structures—the Malpighi layer in the skin and the Malpighian corpuscles in the kidney and spleen.

Soredia and their counterparts, isidia, are not alone in their role as agents for dispersal. They are the most widely used vehicles for vegetative reproduction, but lichens didn't colonize the entire planet by limiting themselves. There are other kinds of propagules (blastidia, phyllidia, schizidia, lobules) and simply broken bits of thallus, especially from the long threads of beard lichens. Even tiny black granules on the lower surface of the frosted rock tripe *Umbilicaria americana* are able to become a new lichen. More things, Horatio.

See also apothecium; isidium.

Substrates

Lichens attach to almost anything. Common substrates are tree trunks, twigs and branches, bare wood, rocks, and soil; less common are bones, mosses, leaves (mainly in the tropics, because temperate zone leaves are too short-lived), other lichens, and even living animals. Non-natural surfaces like glass, plastic, metal, concrete, asphalt, fabric, rope, rubber, and probably more also provide opportunities for attachment. For example, lichens are growing on the abandoned ghost trains in the Allagash Wilderness in Maine.

The actual material of the substrate is only one of the determinants for suitability for any given lichen. Other factors are acidity, chemistry, surface texture, stability, and the persistence of the surface. (Will it stay around long enough for the lichen to first get established and then to grow?) Environmental conditions like humidity, pollution, and amount of light are also significant. Most lichens are quite picky about what they will attach to. For example, the mortar-rim lichen, *Myriolecis dispersa*, requires a calcium-rich substrate but doesn't care whether it's natural (calcareous rocks) or artificial (mortar or concrete). On the other hand, the brightly colored red-and-green Christmas lichen, *Herpothallon rubrocincta*, depends more on climate than substrate. It needs humid tropical to subtropical woodlands but doesn't care whether the trees are hardwood or conifer. It has been found on bald cypress, live oak, magnolia, holly, bay, long-leaf pine, palmetto, and even on mortar. A very few species, called vagrants, have no substrate.

They form balls and roll around in sandy, gravelly dry areas.

All the common substrates have variations that affect their suitability. Tree trunks like beech have smooth bark; others are coarse and creviced; pines and spruces are acidic. Similarly, bare wood may be hard and dry like driftwood or soft and crumbly like rotting wood. Rock substrates tend to be one of two major types: siliceous, like granite or sandstone, and calcareous (calcium-rich), like marble or limestone. By and large a lichen needs not only the right substrate but also the right ecosystem—desert, coast, old-growth forest, second-growth forest, alpine above the treeline, and so on. Within those ecosystems, there are microhabitats where conditions vary widely even across a distance of just a few inches. For example, sulphur dust lichen (*Psilolechia lucida*) grows only in crevices or on the underside of overhangs on rock walls, where the light is lower and the humidity higher than on the top of the wall, which is home to other species.

The most likely animal substrate is probably bone. Classic examples are skulls in the desert and whale-bone on the shore. Living animal substrates include barnacles, the moving boulder otherwise known as the Galápagos tortoise, and some New Guinea weevils with sticky backs.

When you come upon an abandoned car in the back of a junkyard or behind a deserted house, you may well discover lichens on the metal exterior, windshield, and headlights. If the windows are broken or the roof is rusted through, lichens will also be on the dashboard, steering wheel, and upholstery, whether cloth or leather. The bottom line is that if something stays still for long enough, some kind of lichen will grow on it. Keep moving.

See also Galápagos tortoise; lichenicolous fungi; vagrants.

Sunburst Lichens

These aptly named lichens are the bright orange foliose ones you can't help noticing on trees and rocks. Some, like the maritime sunburst lichen seen on rocks on the Atlantic and Pacific coasts and the Gulf of Mexico, are obvious even from a distance. The elegant sunburst lichen can be seen glowing like a halo on the top edge of tombstones in open, sunny graveyards. These are examples of some of their very conspicuous habitats. They grow all around the world, mostly in open habitats. The orange color on the famous prehistoric nuraghes in Sardinia is due to sunburst lichens. When they grow in more shaded spots, the color of the thallus is a paler orange with greenish tones. They are hardy enough to grow in Antarctica and have been found in the Himalayas at around 23,000 feet.

Most species belong to one of three genera, *Xanthoria*, *Xanthomendoza*, or *Rusavskia*. They are fairly easily distinguished from other orange lichens by their foliose growth habit. Firedot lichens in the genus *Caloplaca* have a crustose growth form, and orange bush lichens in the genus *Teloschistes* are fruticose.

Sunburst lichen's orange color is due to the compound parietin in the cortex or outer coating of the lichen. Parietin absorbs blue light, reducing the amount that reaches the photosynthesizing algae in the lichen and thereby protecting it. Too much blue light would damage the algae. The complementary color, orange, is reflected. In shadier habitats the lichen produces less parietin because there is less need for light protection, which is why those thalli are not so boldly orange.

Parietin, an anthraquinone, yields a purple color with potassium hydroxide (KOH), which is one of the common reagents used for spot tests in lichen identification. In this case, with the parietin being in the cortex, a drop of colorless KOH on the surface of a sunburst lichen will immediately turn purple. Parietin has demonstrated antifungal activity against some agricultural mildews and has also shown potential in treating human cancers by slowing the growth rate of cancer cell lines.

Sunburst lichens, especially elegant sunburst, thrive in habitats rich with nutrients from bird and animal droppings. The urea breaks down and nitrogen becomes available, so these lichens flourish on the top of gravestones where birds have been sitting and on rocky ledges used as latrines near crevices and burrows where pack rats or other animals den. The bright color that protects the lichens from excessive light does not do any favors for the animals providing the nutrients. Inuit hunters use the color to locate the homes of hoary marmots more easily, and poachers use it to locate the nests of peregrine falcons. Unintended consequences.

See also crustose; Doctrine of Signatures; foliose; fruticose.

Survival Food

Lichens are not a preferred food for most humans most of the time. They are bitter and hard to digest. Their desirability—I use the term loosely—changes when the other choices are starvation or cannibalism. Even then, they are sometimes second choice. For people in dire straits, lichens have the advantage of being found in places inhospitable to humans, where there may be few other choices, and they are present in all seasons. They provide an option to humans who due to misadventure, plane crashes, war, famine, or other emergencies find themselves in need.

Members of Sir John Franklin's expedition in the Canadian arctic faced starvation in 1821. Those who survived had supplemented their meager diet with different species of lichens collectively known as rock tripe (or in Canada, *tripe de roche*). Some members of his party suffered severe bowel complaints that they attributed to the lichen (even though they were also eating shoe leather and rancid bone marrow). If they had known to keep from eating rock tripe raw and to add wood ash to the water when cooking it, they would have neutralized the lichen acids, thereby reducing the bitterness and some of the side effects. On at least one occasion, they also ate another lichen inappropriately named Iceland moss.

The expedition of Meriwether Lewis and William Clark in the northwestern United States (1804–1806) owed its survival, in part, to Native American Plateau tribes who shared their knowledge about roots and black tree-hair lichen (*Bryoria fremontii*), which provided much-needed sustenance.

More recently, in 1972, Martin Hartwell survived for thirty-two days in the Canadian arctic after crashing his plane while piloting a medevac flight. The only other survivor, David Kootook, a fourteen-year-old boy, died just two days before rescuers arrived. Hartwell had broken both ankles in the crash, but Kootook was able to collect lichens (of unknown type), which they boiled in a concoction with their last scraps of meat. They ate their remaining raisins to kill the taste of the lichens. After Kootook died, Hartwell made a pair of spruce crutches and was able to reach a tree with some lichens, which he scraped off and boiled for a soup. He was not physically capable of collecting enough lichens for survival, and there were three frozen bodies within easy access. The rescue party found evidence of cannibalism.

Stories of lichens as wartime survival food date back to Alexander the Great's army in 330–327 BCE, when they ate "manna" to survive. There is a story that, in the American Revolutionary War, George Washington's troops made rock tripe soup during their ordeal in the winter of 1777–1778 at Valley Forge, Pennsylvania. Much more recently, in the Bosnia-Herzegovina war of 1992–1995, soldiers, guerrilla fighters, and blockaded civilians had to forage for food. In the forests around the Drina River canyon, people collected and ate mushrooms and lichens not normally a part of their diet. They used seven different species of lichens, sometimes in soups and stews and other times ground into flour. The two most common lichens used were oakmoss lichen (*Evernia prunastri*) and straw beard lichen (*Usnea barbata*).

Part of the tradition of indigenous peoples in western Montana, Oregon, British Columbia, parts of Scandinavia, Labrador, and probably other northern latitudes is the use of lichens as famine food when hunting fails, when severe weather eliminates normal sources, or when food is lost by fire or raids. There was a lot of variation among the groups. For example, some of the Salish people in Washington and Montana cooked black tree-hair lichen (*Bryoria fremontii*) with other ingredients to make a luxury food, while others would eat it only in time of famine.

See also food (for humans); Franklin's expedition; manna.

Symbiosis

The term "symbiosis" comes from the Greek *syn* for "together" and *bios* for "life," i.e., "living together." In a broad sense, it could mean everything on planet Earth, but it started out with a singular application. Lichens! The word was coined in 1877 by Albert Bernhard Frank, a German botanist and plant pathologist, in response to the discovery of the dual nature of lichens. He defined symbiosis as "where two species live on or in one another." Anton de Bary, a German contemporary of Frank's and also a botanist and microbiologist, refined the definition. By his criteria, not only must the two species live together, but they must also be in intimate contact and of different species.

The word "symbiosis" is still thought of by many as a mutually beneficial relationship, but that is only part of the story. The word covers a range of situations from parasitism to mutualism, all of which are governed by the extent to which the partners derive benefits from the relationship—or not. In parasitism, one partner benefits at the expense of the other. In mutualism, both benefit. In commensalism, one party benefits and the other neither suffers harm nor gains any benefit. These subsets are convenient but hard to delineate, and sometimes fluid.

Lichens are largely considered mutualists. They are a mini-ecosystem of unrelated organisms in which the mycobiont (the primary fungus) needs carbohydrates for nourishment, and the photobiont (alga, cyanobacterium, or both), which makes carbs, needs protection. These two partners have a supporting cast of other fungi, algae, and bacteria. Both partners benefit. Some lichenologists, however, are dubious. They see the partnership as controlled parasitism in which the mycobiont exploits the photobiont. Trevor Goward, Co-Curator of Lichens at the Beaty Biodiversity Museum at the University of British Columbia, has famously said, "Lichens are fungi that have discovered agriculture." Some crops, such as the green alga (*Trebouxia*), have become so domesticated by lichenizing fungi that they are seldom seen wild. Is this an evolutionary window into fungi developing chloroplasts?

Symbiosis is also categorized by whether the species can survive without each other (facultative) or not (obligate). Yellow-billed oxpeckers in the Serengeti perch on giraffes and pick through their hair for ticks (and other morsels). Sometimes the oxpeckers sleep over in the giraffe's armpits (leg pits?), but they come and go, being free to do so. They also associate with buffalo, zebra, and other large mammals. In other words, the symbiosis between oxpeckers and large mammals is facultative. Compare that with the fungi and algae in lichens. The fungal hyphae wrap the algae all around and contain them. They are committed to living together "till death do them part."

Lichen symbiosis qualifies as obligate mutualism but is deserving of a whole new category. The relationships between oxpeckers and giraffes, between bees and flowers, and so on, allow the partners to retain their own look and feel. They don't become a different thing. The lichen partnership of a fungus and photobiont results in a new entity—a lichen—that doesn't look like either of the partners. The fungus on its own, kept alive in a lab, would be more or less amorphous. When teamed with a photobiont, it takes on a very particular shape and structure. It has taken symbiosis into new territory. The prefix "ultra-" means "beyond" or "on the far side of," so I suggest "ultra-symbiosis" for this phenomenon.

However lichen symbiosis is labeled, it is highly successful. It has enabled lichens to live all over the globe in places where neither fungi nor algae could live on their own.

Tardigrades

Water bear and moss piglet are common names for tiny animals that live on lichens and mosses and in other damp places. Despite having eight legs, they are not classified as arachnids with more familiar eight-legged creatures (spiders, ticks, mites, and so on). They have their own phylum, *Tardigrada*. The name comes from *tardi* meaning "slow" and *grado* meaning "walker." It's descriptive because, despite living in water, tardigrades don't swim; they stumble around slowly with a lumbering gait. (Among the many hundreds of species, there are a few speedy swimmers. The ones you find on lichens are the slow kind.)

Tardigrades can be found from the tropics to the poles. Around 10 percent of the 1,300 species are marine;

the remainder are freshwater or terrestrial, but the latter need at least a film of water in order to function. Imagine a dachshund with no ears, no tail, and two extra pairs of legs. That will give you an idea of the proportions of a tardigrade. Then trade the hairy coat for a cuticle, like an insect exoskeleton, let the toenails grow, and shrink the whole thing down to no more than 1.2 millimeters in length. Now you have something approximating a tardigrade—in its active state. They are somehow endearing.

Any creature that lives on a lichen has to deal with the same cycles of desiccation and rehydration that are typical for lichens. Tardigrades are expert survivalists. In the absence of water, they dry up into a lumpy ball called a tun. In this near-death state, when their metabolism is approximately 0.01 percent of normal, they are able to survive temperatures close to absolute zero or up to 300°F, X-ray radiation at 1,000 times the lethal exposure

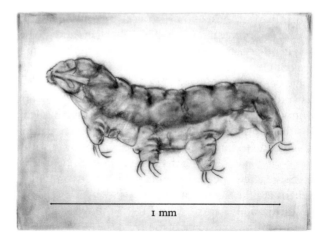

1 mm

for humans, low pressure as in a vacuum, high pressure as much as six times the deepest part of the ocean, lack of water, and lack of oxygen. When conditions are favorable, they revive. The tun stage is viable for decades. There is a case of a dried moss specimen in a museum collection yielding a live tardigrade after 100 years. Despite longevity in survival mode, a tardigrade's life in active mode is as short as a few weeks.

A side benefit of looking closely at lichens is that, once in a while, you will happen upon a tardigrade. Most lichen microscopy is done with dried specimens, in which case a tardigrade would be in its tun state and hard to recognize. You would need to rehydrate it. If you are on a serious tardigrade hunt, you might raise your chances with moss, which has more protected nooks and crannies. If you are a lichen purist, choose a foliose lichen with lots of texture, put the sample face down in some water for a day, squeeze out the sample, transfer the water to a petri dish (or other shallow container), and over the next couple of days scan it at 20–40x under a dissecting microscope. You will see other lichen fauna, like rotifers and nematodes, but will easily recognize the roly-poly body with short legs that is the hallmark of a tardigrade. It is quite possible you will see more than one species.

See also BIOMEX; poikilohydry.

Thallus (*plural* Thalli)

All the fungal, algal, and cyanobacterial components of the lichen, excluding fruiting bodies, constitute the lichen thallus. *Non*-lichenized fungi are typically hidden in soil or wood; we know they are there only when they produce a fruiting body, a mushroom. Lichenizing fungi don't grow like other fungi, even though 90 to 95 percent of the lichen mass is fungal. They transmogrify. The partnership of the components results in a thallus, which is readily visible wherever it grows, whether fruiting or not. Of course, there are exceptions: endolithic lichens, for example, grow in a rock and are noticed when they fruit.

Both because lichens are very old in the context of the tree of life, and because they didn't all evolve at the same time from a common ancestor, it's not surprising that they have enormous variety and exhibit different growth forms. Sadly for those who like things neat and tidy, and happily for those who color outside the lines, many lichens defy rigid categorization on the basis of growth type. The neat and tidy part is covered by three main types: crustose (crusty), foliose (leaf-like), and fruticose (shrubby). The thallus of the nonconformists and those sitting on the proverbial fence might exhibit one type initially, called the primary thallus, and transition over time into another type, the secondary thallus. Others may be crustose in one part of the thallus and foliose in another part at the same time. Some crustose thalli are very thin; in the absence of an upper cortex, they appear as a coating of grainy particles on the substrate. These thalli are described as leprose.

Some lichenologists allow for four major thallus types, the fourth one being squamulose. The word is derived from three components: first the root *squam*, meaning "a scale," then *ule*, meaning "small" (as in "globule," a small glob), and finally *-osus* to indicate "an abundance." Squamulose lichens therefore have a thallus made up of an abundance of small scales. Individual scales are usually attached along one edge (think of cedar shingles on a house). The unattached edges may be smooth or divided. Scale size matters. If the scales are extremely tiny (smaller than one millimeter), they might be grouped with the crustose lichens. If the scales are large (over fifteen millimeters), they could be confused with foliose lichens. Making identification yet more difficult is that some squamulose lichens lose most of their squamules over time, leaving only erect stalks bearing the reproductive structures. At this stage of their life cycle, all that's left is a secondary thallus exhibiting a fruticose form.

Despite somewhat arbitrary boundaries between some of the growth forms, a lichen thallus is usually recognizable as a lichen and seldom confused with other fungi, mosses, or liverworts.

See also crustose; foliose; fruticose.

Theophrastus (ca. 371–287 BCE)

Theophrastus was not just another ancient Greek philosopher, but the first person (we know of) to use the word "lichen" to describe a lichen. (Hippocrates [460–370 BCE] had already used the word for skin diseases involving eruptions of papules, or little bumps on the skin.) Theophrastus was one of Aristotle's students and subsequently a colleague. As a proponent of the Aristotelian system, he furthered its teachings when he took over leadership of the Lyceum, the school founded by Aristotle.

Theophrastus's botanical works stand as significant contributions to early natural history, especially the ten-volume *Enquiry into Plants* (*Historia Plantarum*), of which nine volumes survive. He started it when he was about twenty years old and continued revising it all his life. In it he named over 550 plants, both wild and cultivated, along with everything he could find out about them, including medicinal uses. He supported this work with *Plant Explanations*, a six-volume work in which he methodically accounted for the common or distinctive characteristics of plants. In recognition of those works, he has been called the father of botany. (Lichens would be considered plants until the late nineteenth century.)

It's not at all clear exactly what Theophrastus was referring to when he used the term "lichen." Lichens were obviously encountered by the naturalists of his time, who would have seen them as distinctly different from other plants. They typically grouped lichens with mosses, algae, and liverworts. It is generally agreed that the lichens Theophrastus described (less than perfectly) are in fact what we know today as lichens. Two of them were possibly *Roccella tinctoria* and *Usnea barbata*. There is no certainty on the species.

Of his many works on wide-ranging topics (logic, metaphysics, psychology, ethics, politics, music, human physiology, rhetoric, and botany), few remain, but the two major works on plants are among those that do. They were influential to other botanists up through the Middle Ages. An indication of their importance is the printing of their Latin translations in 1483, not too long after Gutenberg's invention of the printing press around 1436.

See also Pliny the Elder.

Thoreau, Henry David (1817–1862)

Naturalist, essayist, poet, and leading transcendentalist, Henry David Thoreau often had what he chose to call a lichen day, a good lichen day, or somewhat of a lichen day. Such a day would be February 7, 1859, "a warm and moist or misty day in winter." Lichens look lush and vibrant when they are moist. The fungal cells in the cortex are opaque and dull when dry but become transparent when moist, letting the color of the algae come through. The shimmer of the moisture, the plumpness of the thallus, and the richer color make lichens shine. As Thoreau wrote in his journal:

> a little moisture, a fog, or rain, or melted snow makes his wilderness to blossom like the rose . . . when I see the sulphur lichens on the rails brightening with the moisture I feel like studying them again as a relisher or tonic, to make life go down and digest well, as we use pepper and vinegar and salads. . . . To study lichens is to get a taste of earth and health, to go gnawing the rails and rocks.

On a cloudy, misty January 7, 1855, he wrote:

> The bank is tinged with a most delicate pink or bright flesh-color—where the *Baeomyces roseus* [renamed *Dibaeis baeomyces*, but the common name, pink earth, remains the same] grows. It is a lichen day. The ground is covered with cetrariae, etc., under the pines. How full of life and of eyes is the damp bark! It would not be worth the while to die and leave all this life behind one.

In winter, when so much is dormant, Thoreau observed, "A lichenist fats where others starve. His provender never fails."

Tree cattle

These invertebrates are also known, equally incorrectly, as bark lice. The little critters, about one-quarter of an inch long, are obviously not cattle, nor do they look like them. They are members of the family *Psocidae*, commonly called Psocids. ("So, Syd, how are things?") They are not lice, which are parasites and have an entirely different lifestyle, but they are found on tree bark. So both names, tree cattle and bark lice, are half right. I prefer "tree cattle" because it doesn't have the negative association with lice and because these invertebrates hang together and move like a herd. The herding behavior is common to both nymphs and adults. When disturbed, they don't scatter; they move *en masse*, like a river of sheep managed by competent border collies. Tree sheep?

There are two reasons for their appearance in a lichen book. One is that they eat lichen—as well as other organic material such as fungi and algae that they find on bark. Tree cattle do no harm to the tree and could be thought of as bark groomers. The other reason is that anyone who studies lichens is bound to encounter them. If you bring lichen-laden branches into the house for further study and leave them on the dining table overnight, you may find a herd of tree cattle in the morning. This I know from experience. No worries. They're easy to herd onto another branch ("Git along, little dogies") and take back outside.

Tree cattle are members of a very large family. There are seventy-five species in the United States. My dining table visitors were *Cerastipsocus venosus*. One species, known from Scotland, showed such a voracious appetite for lichens that it completely ate away the lichen cortex, which exposed the white medulla, leaving the tree trunks in an area of more than an acre white and ghostly and killing the lichen. More than the icing off the cake, they ate the shell off the turtle. A more discerning tree cattle species just eats the apothecia.

Tuckerman, Edward (1817–1886)

Edward Tuckerman is the man whose name was given to a ravine on the eastern side of Mount Washington in the Presidential Range of the White Mountains of New Hampshire. Tuckerman Ravine is a well-known destination for hikers and expert, adventurous spring skiers. Less well known is the fact that Tuckerman was a serious lichenologist and the first person to investigate the lichen flora of the White Mountains.

Tuckerman did his early work at a time when the dual nature of lichens was not known. One theory suggested that they were mosses that had experienced some developmental disruption. Another maintained that they were formed spontaneously, without seed, from decomposing water (whatever that is) in which vegetable matter had formed in the presence of warmth and sunlight. This vegetable matter was thought to be the common ancestor of algae, mosses, and lichens. It was not until 1866 that the German mycologist Heinrich Anton de Bary suggested that a lichen was not a single organism but two. A radical notion for its time. Then in 1867, Simon Schwendener, a Swiss botanist, published his description of lichens as an intimate algo-fungal union. This was not accepted by all botanists. In the *Biographical Memoir of Edward Tuckerman*, written by W. G. Farlow (for whom the Farlow Herbarium at Harvard is named) in 1887, Farlow states:

> The botanical world was divided in opinion and for the last 15 years the so-called algo-fungal theory of lichens has given rise to endless controversies of a personal and very acrimonious character. . . . It is said that at first he [Tuckerman] was inclined to favor the theory, but if so he soon changed his views and sided with the opponents of the theory. It must be said to his credit that his references to the subject were always courteous and dignified in marked contrast with the course of some other well-known lichenologists.

In his *A Text-book of General Lichenology*, published in 1897, Albert Schneider writes in a less courteous tone about Tuckerman's publication of *Genera Lichenum*, saying that it "is, however, unsatisfactory, because the author did not seem to have any clear conception of genera." (The German word *Schneider* translates to "cutter" in English—just saying.) Schwendener's followers agreed to classify lichens on the basis of the fungal partner, which is still the case.

Tuckerman's rejection of the theory did not stop him from recognizing lichens, collecting them, and describing species not previously known. He was respected by many, and his name lives on in the genera named for him—*Tuckermannopsis* and *Tuckermanella*—as well as species such as *Platismatia tuckermanii* and others. The genus name *Tuckermannopsis* is curious. The suffix "-opsis" means "similar to" or "resembling." The way in which the lichen resembles Tuckerman is just another lichen puzzle—one perhaps for historians rather than lichenologists.

It is entirely fitting that a man who not only remained pleasant in the midst of very challenging academic terrain but also advanced our knowledge of lichen flora should have a challenging and beautiful geological ravine named for him.

See also Schwendener.

Vagrants

Most lichens are attached to a substrate. Some are very tightly bonded, but even with those more loosely attached, a disturbance is likely to break off pieces rather than dislodge the entire lichen. A very few lichen species never attach to anything. They are called vagrants or vagabonds, terms that carry a sense of an entity that makes no contribution to society. Germans called them *Wanderflechten* (the German word for lichen is *Flechten*), which seems less pejorative.

Their abode, to which they are not fixed, tends to be open, dry, sandy ground with little vegetation. By curling up when dry, vagrants shield the photobiont layer near their upper surface from excess ultraviolet radiation and from being sandblasted in strong winds. Some species form relatively tight balls and are easily blown along the ground. They are dispersed by wind, rain, and animals.

The manna of the Israelites is thought to have been balls of vagrant lichen picked up in whirlwinds and dropped elsewhere, as if from heaven. The lichen responsible for the death of elk in Wyoming in 2004 was a vagrant species.

See also manna; substrate; Wyoming.

Wolf Lichen (*Letharia vulpina*)

Wolf lichen owes its common name to Scandinavian lore about its use in poisoning wolves. The genus name *Letharia* comes from the Latin meaning "very deadly" and refers to the lichen's toxic component, vulpinic acid. The species epithet, *vulpina*, comes from *vulpes*, the Latin for "fox"! This epithet isn't as confused as it may seem because the lichen was also used to kill foxes. In fact, the name of the closely related *L. lupina* is based on *lupus*, the Latin for "wolf." The species are difficult to tell apart in the field, but both contain vulpinic acid, which is poisonous to carnivores, including man. Rabbits, mice, and hedgehogs can survive it.

Wolf lichen's use as a poison dates back to a record from 1673 that people whose livelihood depended on reindeer and other livestock were crumbling it and mixing it into bait for trapping foxes and wolves. According to one record, the powdered lichen was cooked with fat and bits of meat, then mixed with "fresh blood and bits of reindeer cheese, so that it smelled good." The poisonous concoction was either pressed between the skin and meat of a reindeer carcass or inserted into the meat. There are also records of wolf lichen being mixed with ground glass. While the predator-control practice was not widespread, knowledge of it was handed down until the early to mid-twentieth century, by which time it had been replaced with strychnine and shotguns.

In northern California, the Achomawai people prepared arrow points for squirrel hunting by leaving them in sodden wolf lichen for a year and enhancing them with rattlesnake venom.

Wolf lichen also had widespread use as a bright yellow dye. Scandinavians dyed wool, Karuk people on the Klamath River in California and the northern Cheyenne of Montana dyed porcupine quills, and other American tribes dyed basketry materials. The Apache used it as face paint for a charm and painted a cross on their feet for super-stealth enabling them to pass their enemies unseen.

Poisonous substances are often medicinal if the dose and the application are properly understood. The Sámi people of Lapland warned against inhaling powdered lichen when preparing wolf poison lest it trigger a nosebleed. And yet it has been used externally by American tribes to dry up running sores and treat a variety of other skin conditions. Taken internally, wolf lichen was

Klamath Modoc basket using wolf lichen dye
Source basket for this drawing courtesy of
The Hallie Ford Museum of Art, Salem, Oregon

a treatment for stomach disorders (ulcers) among the Blackfoot, and the Salish used it for "internal problems."

Its history is as colorful as the vivid yellow-green lichen itself, which has a shrubby growth habit. It is found in the western United States and northern Europe growing on the bark of conifers and birch and also on bare wood like old fences. Linnaeus reported it as common on church roofs and old wooden walls in his home province in southern Sweden. Today in Sweden wolf lichen is rare enough to be red-listed.

Women in the Canopy

Most lichen species that are found in the canopy also occur at low elevations if light, humidity, and substrate are suitable. A specialist lichen, the tiny stubble *Calicium sequoiae*, has only one habitat—the bark of old-growth coast redwoods at heights of 60 to 260 feet. Female scientists are also found mainly at lower elevations but again, there are specialists.

Nalini Nadkarni is a tree climber whose canopy research extends to all the life forms up there and their relationships, of which little was known when she started her work in the 1980s. An integrator of science, arts, and spirituality, she is also a scientist, a conservationist, and cofounder of the International Canopy Network. Nadkarni is a person you wish you had known about when you were young. She shares her passion for trees and their science wherever she can—schools, religious groups, even solitary confinement units in prisons. The lichen *Porina nadkarniae* is named for her.

Margaret Lowman, who came to be known as Canopy Meg, started with research on leaves and insect herbivory in Australian rainforest canopies. She pioneered the use of walkways for canopy access, having tried cherry pickers, cranes, hot-air balloons, and of course climbing. Her influence led to the building in 1987 of Tree Top Walk—thought to be the world's first canopy walkway—in Lamington National Park in Australia and to the first in North America in Hopkins Forest in Massachusetts in 1991. Walkways are now available on all continents (except Antarctica) for both science and eco-tourism.

Marie Antoine started climbing trees as a preschooler. In a sense, she has never come down. Her college research into lettuce lichen, *Lobaria oregana*, took her into the tall trees of the Pacific Northwest. She now works with a team on analyzing the role of the biggest, oldest trees in ameliorating climate change and ensuring there will be old-growth forests in the future. When not carrying out research in the unique habitat of the country's tallest trees (Sitka spruce, coast redwood, giant sequoia, and Douglas fir), Antoine lectures in the Department of Biological Sciences at Humboldt State University in California. The only access option for her canopy work is climbing.

These women were pioneers in a male-dominated arena. They are an inspiration for female canopy ecologists in the future. The lonely, red-listed *Calicium sequoiae* lichen may have roommates waiting for recognition.

Wordplay

Mycologists and lichenologists have all heard: "I'm a fun guy. Won't you take a lichen to me?" I have a T-shirt adorned with images of jelly lichens (*Collema* genus) along with the text: "How you lichen me now?" The field of wordplay is broad and the surface of lichen territory has barely been scratched. There're many more possibilities than lichen/liken. Yes, it's a minefield of groan-worthy puns, but worth exploring for intrepid word lovers.

The old word "lycanthrope," meaning a type of werewolf, has been adopted into the modern horror genre and abbreviated to "lycan." Conversation overheard at a movie location: "Is this the place where wolf lichens grow?" asked the lycan. "Yes, look right here. They're all over my family tree," replied Li Chen, the Chinese actor and director, aka Jerry Li.

Did KFC marketeers realize that their slogan "Finger lichen good" was paying homage to etymology? The word "lichen" originated from the ancient Greek word λειχήν, meaning "the licker," which was derived from the verb λείχω, "to lick." It eventually became "lichen," the word we use today, but it's nice to find a corporate entity honoring the original meaning.

The clever pun "Ways of Enlichenment" is worthy of mention because it is the name of an excellent website, blog, and book featuring the lichen knowledge of the enlightened Trevor Goward.

"I contend that lichens are really plants!" said Tom, speciously.

See also etymology.

Wyoming

The mysterious deaths of large numbers of elk in south-central Wyoming in 2004 turned out to be caused by a lichen. Elk normally eat lichen in winter, so lichen was not initially suspected. After wildlife veterinarians ruled out other causes, they enlisted the help of lichenologists. Emergency room doctors ask mycologists to examine stomach contents pumped from victims of suspected mushroom poisoning—but why should mycologists have all the fun?

James Lendemer, Staff Lichenologist and Associate Curator of the Institute of Systematic Botany, New York Botanical Garden—and in this case, lichen detective and dung delver—was the lucky recipient of fecal samples. He was able to detect lichen substances and confirm that the tumbleweed shield lichen (*Xanthoparmelia chlorochroa*) had passed through the animal's gut. The wildlife vets set up an experiment. They fed the suspect lichen to three healthy elk, who all succumbed, but the mystery remained. Most elk in the area didn't die, yet they all ate the lichen. Grazing animals like cows and sheep and wild animals like deer, antelope, and caribou depend on microorganisms in their gut to break down their food. The theory in Wyoming is that the susceptible elk were migrants from Colorado with a different gut microbiome that was not able to neutralize the toxic lichen acids.

Perhaps providing more evidence for the microbiome theory is that pronghorn antelope (a different species with different gut microbes) eat the same lichen with no ill effects. The pronghorn, with its symbiotic microscopic helpers, overcome the fungus-alga partnership. Mutualism versus mutualism. Wildlife managers have even used the presence of *Xanthoparmelia chlorochroa* to assess habitat suitability before reintroducing pronghorns to areas in the Great Basin.

An older report, also from the western United States, warned that the same lichen (at that time called *Parmelia molliuscula*) contained enough selenium to seriously affect sheep and cattle.

See also lichen substances.

Yosemite

Steep rock faces in Yosemite National Park are blackened by lichens. Just as the human imagination sees "the man in the moon" in the shadows of the moon's landscape, we can see tearstained cheeks in the dark trails of lichens on Yosemite's rock faces. Two of the giant stone structures in Yosemite are called Half Dome and Washington Column. The Native American name for the Half Dome is Face of a Young Woman Stained with Tears because, in the legend of Tis-sa-ack,

she and her husband were punished by the Great Spirit and turned into the stone structures of Half Dome and Washington Column. The dark drip lines of her tears are made of crustose lichens of several species, one of which is the brown tile lichen, *Lecidea atrobrunnea*, which is as long-lived as the legend.

Zone of Inhibition

When a colony of bacteria on an agar plate is not able to grow in the vicinity of a test substance, the bacteria-free area is called the "zone of inhibition." The size of the zone indicates the strength of the test substance. Indigenous peoples around the world have a long tradition of applying lichens to burns, sores, and other wounds to prevent infection. Scientists have confirmed the antibacterial properties of various lichen substances against certain bacteria by observing and measuring zones of inhibition.

For example, they have shown that usnic acid, present in many lichens, including old man's beard, interferes with RNA and DNA synthesis in pathogenic, gram-positive bacteria such as *Staphylococcus aureus*. Other techniques have proven that lichen extracts from some species of shield lichen and lungwort are active against *Mycobacterium tuberculosis*. The work goes on as drug-resistant strains of bacteria become more of a threat.

Zoochory

Dispersal of plants by animals is called "zoochory," pronounced zoh-uh-kory. The term has carried over into the lichen lexicon, because lichens used to be considered plants.

Many lichens have vegetative propagules designed to be easily dislodged by abrasion from wind, water, or animals. Animals that share the habitat accidentally disperse these pieces by walking over the lichens, grazing them, or brushing against them. Some animals are more deliberate. Birds tear off pieces of lichen or gather lichen-coated sticks to use in their nests, which eventually collapse. Some of the lichen pieces may end up in a place where they can grow. In New Guinea, three species of foliose lichens grow on the backs of moss forest weevils, *Gymnopholus lichenifera*. The primary purpose may be camouflage for the weevil, which transports the lichens to new locations, in spite of being slow-moving and flightless. In the United States, larvae of the lacewing insect *Leucochrysa pavida* provide a similar transport function. The lichens are not growing on the larvae but have been collected by them to use for camouflage.

See also camouflage; dispersal; nests.

Lacewing larva
Leucochrysa pavida

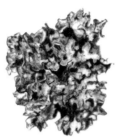

Acknowledgments

For the opportunity to do this book, I thank Robert Kirk and the whole team at Princeton who were a delight to work with, especially Megan Mendonça and Cindy Buck for her copyediting skills and insightful questions. I also thank the two anonymous readers for their valuable suggestions.

For my life with lichens, I thank, now and forever, Elizabeth Kneiper, who taught me lichens, showed me the way, and continues to support my interest. I thank Sue Edwards for her beautiful drawings. I thank the following for content ideas, contributions, comments, editing, and encouragement: Marie Antoine, Imadiel Ariel, Judy Asarkof, Joeth Barlas, Rose Christian, Pete Clark, Andrew Joslin, James Lendemer, Bonnie Miskolczy, Janice Morse, Michaela Schmull, Carolyn Tobin, Ian Watkins, and Mary Zoll. I also acknowledge the Eagle Hill Institute in Steuben, Maine, for offering lichen classes with high-caliber instructors like Troy McMullin and Steve Selva.

Useful References

Allen, Jessica L., James C. Lendemer. *Urban Lichens: A Field Guide for Northeastern North America*, illustrated by Jordan R. Hoffman. New Haven, CT: Yale University Press, 2021.

Bland, John H. *Forests of Lilliput: The Realm of Mosses and Lichens.* Hoboken, NJ: Prentice-Hall, 1971.

Brodo, Irwin M., Sylvia Duran Sharnoff, and Stephen Sharnoff. *Lichens of North America.* New Haven, CT: Yale University Press, 2001.

Casselman, Karen Diadick. *Lichen Dyes: The New Source Book.* Cheverie, Nova Scotia: Studio Vista Publications, 1996; rev. ed., Mineola, NY: Dover Publications, 2001.

Hawksworth, David L., and David J. Hill. *The Lichen-Forming Fungi.* New York: Chapman and Hall, 1984.

Malcolm, Bill, and Nancy Malcolm. *The Forest Carpet: New Zealand's Little-Noticed Forest Plants-Mosses,*

Lichens, Liverworts, Hornworts, Fork-Ferns, and Lyco-pods. Nelson, New Zealand: Craig Potton, 1989.

Nash, Thomas H., III, ed. *Lichen Biology,* 2nd ed. Cambridge: Cambridge University Press.

Richardson, David. *The Vanishing Lichens: Their History, Biology and Importance.* Exeter, UK: David & Charles Ltd., 1975.

Rogers, Robert R. H. *The Fungal Pharmacy: The Complete Guide to Medicinal Mushrooms and Lichens of North America.* Berkeley, CA: North Atlantic Books, 2011.

Schneider, Albert. *A Text-Book of General Lichenology, with Descriptions and Figures of the Genera Occurring in the Northeastern United States.* Binghamton, NY: W. N. Clute & Company, 1897.

Sheldrake, Merlin. *Entangled Life: How Fungi Make Our Worlds, Change Our Minds, and Shape Our Futures.* New York: Random House, 2020. See especially chapter 3 on lichens, "The Intimacy of Strangers."

Stephenson, Steven L. *The Kingdom Fungi: The Biology of Mushrooms, Molds, and Lichens.* Portland, OR: Timber Press, 2010.

Tripp, Erin, and James Lendemer. *Field Guide to the Lichens of Great Smoky Mountains National Park.* Knoxville: University of Tennessee Press, 2020.

Useful Websites

Australian Lichens, https://www.anbg.gov.au/lichen/
   index.html
British Lichen Society, https://britishlichensociety
   .org.uk/
Consortium of Lichen Herbaria, https://lichenportal.
   org/portal/index.php
Enlichenment, https://www.waysofenlichenment.net/
Mycopigments, https://www.mycopigments.com/

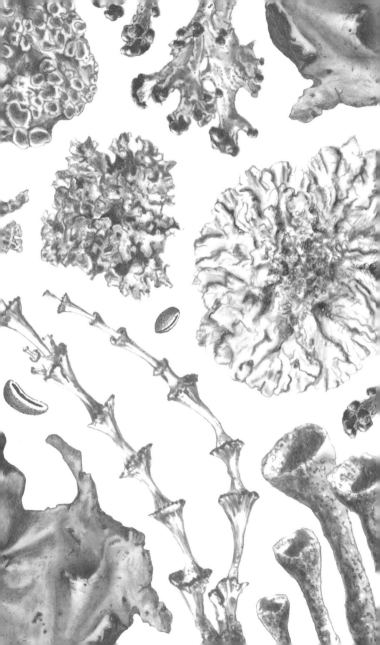